Ionic Liquid Monopropellants for Advanced Propulsion

Drishyam

TABLE OF CONTENTS

CHAPTER 1
INTRODUCTION

1.1 Propellants

Propellants are chemical substances that serve as the core working medium in propulsion systems, providing the necessary thrust to propel rockets, missiles and spacecraft. The thrust is generated by expelling mass at high velocity, typically in the form of hot gases produced through combustion or decomposition of the propellants. Propellants are broadly classified into three main categories: solid, liquid, and hybrid (Mayer et al., 2018; Pradhan et al., 2020) each differing in physical state, storage configuration, and combustion behaviour. Among these, liquid propellants offer significant advantages in controllability and performance, and are further subdivided into monopropellant and bipropellant systems. In monopropellant systems, a single chemical substance serves as both the fuel and the oxidiser. Upon contact with a catalyst or on heating, the propellant undergoes an exothermic reaction, releasing hot gases to generate thrust. This configuration simplifies the propulsion system by requiring only one storage tank and one feed line, making it ideal for spacecraft. In contrast, bipropellant systems involve the separate storage of fuel and oxidizer in two distinct tanks. These components are injected into a combustion chamber where they mix and react energetically to produce high-temperature exhaust gases. While this setup is more complex, it enables higher energy release and greater thrust, making bipropellants the preferred choice for main propulsion in many space launch and interplanetary missions (Remissa et al., 2023). However, monopropellant propulsion systems are widely employed for orbit insertion and adjustment, station-keeping, attitude control, and deorbiting (Pogorielov, 2024) and are inherently reliable and easier to operate than bipropellant or hybrid systems.

Hydrazine (N_2H_4) is by far the most favoured propellant for primary spacecraft propulsion and attitude control thrusters due to its excellent ignition and combustion properties (Sutton, 2001). Hydrazine based propellant have the advantages of long-term storability, simplicity, cold start capability, lower cost and are highly reliable for long term mission requirements (Werling, et al., 2020). Despite these advantages, hydrazine is highly toxic. Its toxicity has led to increasing regulatory restrictions and higher operational costs, particularly during launch pad procedures such as fueling. It is extremely harmful to living tissues and is classified as a probable human carcinogen. Exposure can cause a range of adverse effects, including damage to the liver, kidneys, nervous system, and red blood cells (Choudhary et al., 1997). Hydrazine has a high vapor pressure at room temperature causing it to fume heavily and pose a serious respiratory hazard. As a result, all handling requires the use of Self-Contained Atmospheric Protective Ensemble (SCAPE) suits. These protective measures add significant logistical complexity and often extend payload processing times (Whitmore et al., 2013). Hydrazine has been added to the EU's REACH (Europe's Registration Evaluation Authorization and Restriction of Chemicals) list of Substances of Very High Concern (SVHC), and its future use may be restricted or banned. As a result, several alternative propellants are currently under investigation to identify safer "green" monopropellants (Masse et al., 2013; Werling et al., 2017). A viable hydrazine replacement should offer chemical and thermal stability, a low freezing point, and improved performance. A recent Japan Aerospace Exploration Agency (JAXA) study highlights two key factors for low-cost space access: (i) reduced production, handling, and transport costs due to low toxicity and explosion risk, and (ii) simplified subsystem and interface design.

1.2 Propulsive parameters

Chemical propellants release stored energy by chemical reactions, generating high-temperature gases that expand and accelerate through a nozzle to produce thrust. The thrust (F), depend on the propellant mass flow rate, dm/dt (kg/s) and the average velocity of the exhaust gas, Ve (m/s) is given as (Kamal et al., 2011)

$F= (dm/dt).Ve, \quad or \quad m.\mathrm{Ve}$

The thrust specific impulse (Ns/kg) is given by Isp= F/m and more often the specific impulse is given in seconds, defined as Isp (s)=F/m.g, where g is the gravitational constant (9.8 m/s^2). The density specific impulse is given by Isp. ƒ, where ƒ is the density of the propellant.

1.3 Green Monopropellants

Green propellants are characterized by low toxicity, minimal hazard, and environmental friendliness throughout spacecraft development, launch, and operation. Unlike conventional propellants like hydrazine and its derivatives, which require strict safety protocols and specialized equipment such as SCAPE suits, green propellants offer safer handling and improved storability (Nosseir et al., 2021). Their favourable properties contribute to reduced costs in transportation, storage, and ground operations, enhancing their overall commercial value (Marshall and Deans, 2013).

However green monopropellants face several challenges that limit their wider adoption. Their technology maturity remains low due to limited data on storage behaviour and stability. Most green monopropellants require elevated catalyst bed temperatures to initiate decomposition, which increases system complexity and power demand. Catalyst degradation is another major issue due to the high adiabatic flame temperature. In addition, they are hygroscopic and have material compatibility issues. Thus, although green monopropellants offer significant safety and environmental benefits, these disadvantages currently constrain their widespread implementation in propulsion systems. **Table 1.1** lists the main advantages and disadvantages of green propellants with respect to traditional monopropellant.

Table 1.1 Main advantages and disadvantages of green propellants with respect to traditional monopropellant (Ordonez Valles et al., 2022).

Advantages	Disadvantages
■ Reduced toxicity ■ Reduced safety precautions, no SCAPE suits required ■ Improved efficiency and reduced complexity ■ Reduction in the cost of production, transportation and on-ground operations	■ Not a matured system ■ Less green propellants are flight proven ■ Hygroscopic nature ■ Preheating of catalyst bed required ■ Material compatibility issues ■ High adiabatic flame temperature

Green monopropellants are classified into three major categories: (1) Hydrogen Peroxide Aqueous Solutions (2) Nitrous oxide Monopropellants (either in binary compound, nitro compound, or premixed/blend form) and (3) Energetic Ionic Liquids (EILs) (or premixed oxidizer/fuel ionic aqueous solutions) (Werling, et al., 2020; Gohardani et al., 2014).

1.3.1 Hydrogen peroxide monopropellant

High-Test Peroxide (HTP) is a very concentrated hydrogen peroxide (70-98%). Rocket-grade HTP used in space propulsion for low- and medium-thrust applications is 98% concentration. The non-toxic nature and high density (1.43 g/cm^3) of 98% HTP make it a desirable choice for propulsion system. As a monopropellant 90% hydrogen peroxide decomposes over a solid catalyst bed with the theoretical specific impulse of approximately 154 s. The different classes of hydrogen peroxide and their properties are given in **Table 1.2.** Hydrogen peroxide aqueous solution delivers the lowest performance among the green monopropellants. However, as an oxidiser in a bipropellant with hydrocarbon fuel it can attain Isp exceeding 325 s. The distinctive property of concentrated hydrogen peroxide is its hypergolic ignition with certain fuels, making it suitable for in-space propulsion (Naseem et al., 2017; Pasini, et al., 2015). The development of green hypergolic ionic liquids as new generation of green propellants is reported (Bhosale et al., 2016; Bhosale and Kulkarni, 2017) making use of the hypergolic ignition characteristics of hydrogen peroxide with ionic liquid fuels such as 1-ethyl-3-methyl imidazolium cyanoborohydride ([EMIM][BH_3CN]) and 1-allyl-3-ethyl imidazolium cyanoborohydride ([AEIM] [BH_3CN]).

Table 1.2 Properties of H_2O_2 with different concentrations.

Properties	**H_2O_2 propellant classification**			
Grade	Type 70 (ES)	Type 85 (ES)	Type 90 (ES - HP)	Type 98 (ES)
Concentration %	71-73	85-87	90-91.5	98-99
Density (g/cm^3)	~1.29	~1.34	~1.40	~1.43
Freezing point (°C)	-40	-17	-12	-2

1.3.2 NO_x based monopropellant

Nitrous oxide (N_2O) is the only oxide of nitrogen that falls under green propellant category. N_2O possesses good storability characteristics at room temperature for long term storage when compared to H_2O_2 or cryogenic Liquid Oxygen (LOX). Liquid N_2O has a storage density of ~0.745 g/cm^3 at 20 °C and ~5.2 MPa vapor pressure (Wallbank et al., 2004). N_2O as monopropellant has an Isp of 206 s which is higher than that of high-test peroxide (Zakirov et al., 2000). Nitromethane (CH_3NO_2) is another promising green monopropellant with higher volumetric specific impulse than hydrazine along with a freezing temperature of −28.4 °C.

Nitrous oxide fuel blend used nitrous oxide with ethylene exhibits theoretical vacuum specific impulse of 303 s at an Oxidizer to Fuel (O/F) ratio of 6 under 230 K and 2.5 MPa pressure with a high combustion temperature up to 3264 K (Werling et al., 2017). Nitrous oxide/ethanol blend, with a specific impulse of 331 s and a combustion temperature of 3093 K is also reported (Mayer et al., 2018). Nitrous oxide fuel blends, are nontoxic with low freezing point, specific impulse higher than that of hydrazine and vapor pressure ~50 bar at room temperature, provide self-pressurization capabilities, which allow simple feed-system design. The density and performance of liquid NO_x monopropellants are given in **Table 1.3.**

Table 1.3 Density and performance of liquid NO_x monopropellant (Liu et al., 2018; Nosseir et al., 2021; Remissa et al., 2023).

Propellant	**[a]N_2O (Liquid)**	**[b]Nitromethane**	**[b]NOFBX**	**[d]H_yNO_x(Ethene)**	**[e]NO_x/ Ethanol**
Density (g/cc)	0.745	1.1371	0.700	0.879	0.892
Isp (s)	206	289	350	303	331
[a]Pc=0.3MPa, [b]Pc=1 MPa, [c]Pc=0.7 MPa and stoichiometric O/F=3, [d]Pc=2.5 MPa and stoichiometric O/F=6, [e]Pc=1 MPa and stoichiometric O/F=5.73					

1.3.3 Energetic Ionic liquid monopropellant

Ionic liquids are a unique class of compounds with melting points below 100°C, and have recently been the focus of research due to a wide variety of desirable properties. Energetic ionic liquids (EILs) are a specific class of ionic liquids with nitrogen-rich cations, coupled with oxygen-rich anions such as nitrates, sulfates, and perchlorates. Compared with hydrazine and other monopropellants, energetic ionic liquids offer several practical advantages: they are typically liquid at or near ambient temperature (low melting point), have negligible vapor pressure and lower toxicity. These attributes make them easier and safer to store and handle, while their higher densities often enable competitive or improved performance in monopropellant formulations. Ammonium dinitramide (ADN) and hydroxylammonium nitrate (HAN) are among the promising energetic ionic liquids (EILs) as replacements for hydrazine owing to their improved performances with lower toxicity, higher density, and higher specific impulse than the conventional propellants (Gohardani et al., 2014). However, they have melting points above room temperature, and can be used as a storable liquid propellant in an aqueous solution.

ADN a white solid salt consisting of the ammonium cation (NH_4^+) and the dinitramide anion [$N(NO_2)_2^-$] is developed as an oxidizer for solid rocket propellants by the Soviet Union. ADN exhibits an endothermic melting peak at 91–93 °C, followed by an exothermic decomposition in the range of 150–210 °C, and an endothermic decomposition of ammonium nitrate (AN) in-situ formed during the decomposition of ADN. The TGA curve of ADN indicates a single-stage decomposition with 100% mass loss at 220 °C. ADN has an oxygen balance of 25.8%, density of 1.81 g/cm^3(Andreev et al., 2000; Bottaro et al., 1997; Kumar, 2018; Mishra and Russell, 2002) and high heat of formation of -125.3 kJ mol^{-1} (Wingborg, 2019). ADN as an oxidizer exhibits incompatibility issues with isocyanates in HTPB-based propellants (Landsem et al., 2012). ADN/HTPB- based propellants have low energetics and poor mechanical properties. Consequently, ongoing research in the rocket-propulsion community seeks to identify and optimise fuel binders and plasticizers that are compatible with ADN to achieve maximum energy. The ADN-based monopropellants family mainly consists of FLP-103, 105, 106, 107 and LMP-103S. **Table 1.4** gives the properties of ADN based monopropellants in comparison with that of hydrazine.

Table 1.4 Properties of ADN based monopropellants in comparison to hydrazine (Remissa et al., 2023; Wingborg, et al., 2004).

Propellant	**FLP 103**	**FLP 105**	**FLP 106**	**FLP 107**	**LMP 103**	**Hydrazine**
Formulation	ADN (63.4%)/ Methanol (11.2%)/ Water (25.4%)	ADN (65.7%)/ Fuel[c] (20.7%)/ Water (13.6%)	ADN (64.6%)/ MMF (11.2%)/ Water (25.4%)	ADN (63.4%)/ DMF (11.5%)/ Water (25.3%)	ADN (63.0%)/ Methanol (18.4%)/ Ammonia (18.6%)	-
Density (g/cc)[a]	1.31	1.405	1.357	1.351	1.24	1.004
Isp [b] (s)	254	261	255	258	252	233

[a]Density at 25°C. [b] Chamber pressure=2 MPa and expansion rate =50.
[c]Fuel refer to F-5 as code name

Different fuels are used within this energetic ionic liquid mixture such as methanol, monomethyl-formamide (MMF) and dimethyl-formamide (DMF). However, methanol is found incompatible with ADN unless adding ammonia (NH_3)

to increase the pH of the mixture. LMP-103S based on ADN/water/fuel mixtures has been successfully demonstrated in the PRISMA satellite (Persson et al., 2019). LMP-103S comprised of 63.0% ADN dissolved in methanol (18.4%), water (14.0%) and ammonia (4.6%) has 8% greater specific impulse (253 s) and a 24% higher density than hydrazine, despite being less toxic and noncarcinogenic. The water acts as the solvent for the ADN oxidising agent and methanol fuel component, and it also helps to moderate the reaction temperature. The ammonia serves as a stabilizer during its storage. The advantages and disadvantages of LMP 103S is given in **Table 1.5.**

Table 1.5 Advantages and disadvantages of ADN based LMP 103S propellant (Gotzig Ulrich, 2015).

Advantages	Disadvantages
Non-toxic	Preheating to >250°C required
Low handling cost	High combustion temperature
High density	Requires high temperature resistant chamber material
High performance	
Easy transportation	Friction sensitive

1.3.3.1 Hydroxyl Ammonium Nitrate (HAN)

Hydroxylammonium nitrate (HAN, $NH_3OH{\cdot}NO_3$) is a promising green alternative to hydrazine due to its high energy density and environmentally friendly combustion products. The HAN-based AF-M315E propellant formulation is reported to have a volumetric specific impulse of 391 g s/cm^3 which is higher than that of ADN-based monopropellant (Nosseir et al., 2021). Compared to hydrazine, HAN offers several advantages: higher enthalpy of formation, greater specific impulse, higher density, and significantly lower vapor pressure which reduces toxicity and eliminate carcinogenic risk.

HAN can also serve as an oxidizer in solid composite propellants when combined with fuels, binders, and plasticizers, potentially replacing traditional oxidizers like ammonium perchlorate (AP). However, in solid form, HAN is highly hygroscopic and thermally unstable. It melts at 48 °C and remains stable up to

140 °C, depending on purity. For safety in propulsion applications, HAN is typically used in concentrated aqueous solutions to reduce explosion risk. A 13 molar solution has a specific gravity of 1.523 g/cm³, making it denser than hydrazine (Amrousse et al., 2017; Hoyani et al., 2017; Meng et al., 2009). The boiling and freezing points of aqueous HAN depend on water content. The water content also affects viscosity and specific impulse. The monopropellant's performance is further influenced by the addition of a compatible fuel. Some basic properties of HAN in comparison to those of hydrazine are summarised in **Table 1.6.**

Table. 1.6 Comparison of properties of HAN and hydrazine (Hoyani et al., 2017; Marshall and Deans, 2013).

Parameters	**Hydrazine**	**HAN**
Molecular formula	N_2H_4	$N_2H_4O_4$
Density (kg/m^3)	1003.7	1830
Freezing point (K)	275	243
Boiling point (K)	387	443 (95% HAN)
Heat of formation (kcal/mol)	12.05	-90.3 (95% HAN)
Density Isp (kg/m^3 s)	241	398.67
Chamber temperature (K)	800-900	>1800
Vapor Pressure (Pa)	1970	13.789(95%)
Toxicity (LD50 orally mg/kg)	59	325
Environmental hazard	High, Carcinogenic	Low, Environmentally benign

HAN offers superior performance compared to hydrazine and its high-density impulse enable greater payload capacity and extended mission lifetimes due to efficient storage. Additionally, its low toxicity reduces risks during handling, storage, and combustion, contributing to lower operational costs and producing environmentally benign combustion products, minimizing pollution. The oral median lethal dose (LD_{50}) which is the amount required to cause death in 50% of test animals, indicates that HAN formulations exhibit significantly lower toxicity than hydrazine (Freudenmann et al., 2019). These advantages position HAN as a strong candidate for next-generation satellite propulsion systems (Hoyani et al., 2017; Meng et al., 2009).

Molecular structure of HAN is shown in **Fig. 1.1.** HAN molecule consists of a combustible cation (NH_3OH^+) and an oxidizing anion (NO_3^-), forming an energetic ionic liquid. The redox-sensitive H–N–O group plays a key role in its reactivity. In monopropellant applications, HAN undergoes internal redox reactions, oxidation of the hydroxylammonium cation and reduction of the nitrate anion, which enable self-sustained decomposition without the need for an external oxidizer (Nosseir et al., 2021).

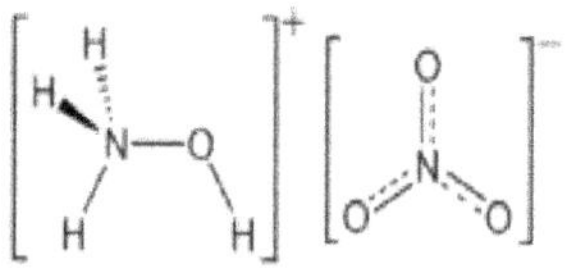

Figure 1.1 Molecular structure of HAN.

A notable feature of HAN is the presence of two hydrogen-bonding functional groups: a strong proton donor (-OH) and a strong hydrogen-bond acceptor (-NH_2). These enable multiple hydrogen-bonding interactions, including O–H···O, N–H···O, and O–H···N types. Density Functional Theory (DFT) (Alavi and Thompson, 2003) studies reveal that the most stable molecular configurations support up to three hydrogen bonds per hydroxylammonium ion as shown in **Fig. 1.2.**

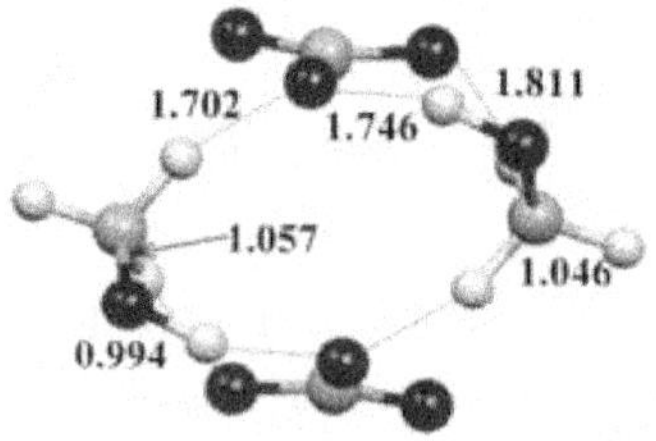

Figure 1.2 Stable molecular configuration of HAN (Alavi and Thompson, 2003).

1.3.3.2 Stability of Hydroxyl Ammonium Nitrate

HAN solutions do not have long-term stability and are highly sensitive to the presence of contaminants, temperature fluctuations, or pH changes. Acidic environments cause autocatalytic decomposition producing ammonium nitrate, nitric acid and dihydroxylamine as by-products. The dihydroxylamine produced is naturally unstable at lower pH and decomposes into water and nitrous oxide (Kasbi et al., 2025).

HAN is incompatible with alkalis, many metals (Reed et al., 2001) and even with compatible materials it undergoes decomposition over time (Schmidt, 1990). Metallic contaminants such as iron and copper have a substantial effect on HAN decomposition reactions. The reductive character of iron and copper ions and their easy production of equivalent redox pairs can cause HAN decomposition. Iron initiates a reaction sequence for HAN that involves the reduction of Fe(III) followed by the oxidation of Fe(II) by nitrate with the generation of nitrous acid. HNO_2 formed further react with HAN, causing the autocatalytic reaction (Harlow, et al., 1998).

$$2NH_3OH^+ + 4Fe^{+3} \rightarrow 4Fe^{+2} + N_2O + H_2O + 6H^+$$

$$2Fe^{+2} + HNO_3 + 2H^+ \rightarrow 2Fe^{+3} + HNO_2 + H_2O$$

$$HNO_2 + NH_3OH^+ \rightarrow N_2O + 2H_2O + H^+$$

The key to the stability of HAN involves storage in low temperature conditions with low Fe and nitric acid impurities and scavenging any nitrous acid formed. The use of chelating agents or stabilizers can mitigate this by sequestering metal ions and reducing decomposition rates (Broemmelsiek et al., 2021).

1.3.3.3 Synthesis Methods of HAN

The purity and performance of the final HAN depend on the synthesis route selected (Hoyani et al., 2017). The synthesis routes for HAN include double decomposition reaction, neutralization of hydroxylamine, reduction of oxides of nitrogen, electrolysis and hydrogenation of nitric acid (Levinthal et al., 1993;

Cawlfield, 1993; Kartte, 1974; Fuchs et al., 1989; Liggett, 1978; Dotson et al., 1989; Hoyani et al., 2017; Yoo et al., 2024). The different synthesis methods are detailed below:

Double decomposition

Double decomposition reaction involving the stoichiometric reaction of hydroxylamine salt and metal nitrate salt in aqueous or organic medium to yield a dilute solution of hydroxylammonium nitrate (Thomas, 1976). The metal nitrate salt used in double decomposition method should have higher solubility in the reaction medium, and the corresponding metal by-product should be insoluble to avoid the presence of impurity in the final product (Hoyani et al., 2017) The metal nitrates precursors can be salts of barium, calcium and sodium ions.

$$(NH_2OH)_2\,H_2SO_4\,(aq) + M(NO_3)_2\,(aq) \rightarrow 2NH_2OH\cdot HNO_3\,(aq) + MSO_4(s)\downarrow$$

Neutralization

HAN can be prepared by a reaction between free hydroxylamine (HA) base and nitric acid (neutralization reaction) in aqueous or organic medium (Clawfield, 1993). The reaction is carried out at low temperature by the slow addition of nitric acid to the HA solution. The reaction is complete when the pH of the solution reaches 2.5. This point can be obtained by calculating the hydrogen ion concentration at the equivalent point by knowing the dissociation constant of hydroxylamine weak acid.

$$NH_2OH\,(aq) + HNO_3\,(aq) \rightarrow NH_2OH\cdot HNO_3\,(aq)$$

In contrast to double decomposition method, neutralization yields HAN without any metal ions impurity.

Electrolysis of nitric acid

Electrolysis of HNO_3 involves electrochemical cell with cathodic and anodic compartments and a separator in between. Aqueous nitric acid is used as a catholyte in cathode made of mercury. The concentration of nitric acid in the two compartments are crucial, since at low concentrations hydroxyl amine is reduced to ammonia at cathode and at high concentration of acid the produced HAN was converted to oxides of nitrogen. The process is limited by the toxicity of the cathode material and non-uniformity in the final product (Dotson, et al., 1989).

Reduction of nitrogen monoxide

Nitrogen monoxide could be reduced catalytically in presence of dilute acid at elevated temperature. Sulfur-poisoned platinum catalyst increases the selectivity of HAN to 86.5 %. This method involves the release of nitrous oxide and the regeneration of catalyst for the process increases the cost and complication of the process (Kartte et al., 1974).

Electrolysis

Hydroxylammonium nitrate is produced from hydroxylammonium sulfate or hydrochloride by an electrolysis method, which uses cation and anion exchange resins and a double decomposition reaction (Wheelwright, 1977). The method is complicated and possesses an explosion risk during reaction between ion-exchange resin and nitric acid, hence unattractive commercially.

Hydrolysis of cyclohexanone oxime

Zhao et al., (2015) reported to synthesis HAN from the hydrolysis reaction of cyclohexanone oxime (CHO) in the presence of nitric acid. The conversion rate and selectivity of the process depend on reactant mol ratio, reaction time, temperature and the extraction solvent used. The conversion of CHO can reach 62.6% with 96.9% selectivity to HAN and 97.4% selectivity to cyclohexanone. The study concludes that the reaction is reversible and the conversion of CHO can be improved by using different extraction solvents.

1.3.3.4 Concentration of aqueous dilute HAN solution

Low concentrations of aqueous hydroxylammonium nitrate cannot be used in propellant applications. Hence, the dilute aqueous HAN obtained during the synthesis need to be concentrated. The dilute HAN solution can be concentrated by evaporation of water in a rotary evaporator under vacuum. The removal of water from HAN becomes more difficult as the HAN concentration increases which makes the process time consuming. Moreover, any rise in temperature during distillation or any air leak to the hot HAN solution results in fuming of the solution leading to the exothermic decomposition of HAN. Dotson and Brooker (1989) describe a process of concentrating hydroxylammonium nitrate solution through the use of membranes. The method involves contacting the dilute salt solution with the sorption side of the membrane to sorb the solvent and permit the solvent to flow through the membrane to the desorption side. This method is attractive but economically unfeasible and relatively slow.

1.3.3.5 Characterisation of HAN

The concentration of HAN in solution can be determined by the titrimetric method (Decker et al., 1988). The titrimetric analysis of HAN has been investigated for the visual endpoint method and potentiometric method. The visual method uses phenolphthalein indicator and alcoholic KOH titrant. The concentration of HAN in presence of fuel such as triethanol ammonium nitrate (TEAN) is carried out by converting HAN to nitric acid by reacting with aldehyde and endpoint is determined using bromophenol blue indicator. Potentiometric titration procedures use an automatic titrator and tetrabutylammonium hydroxide in denatured alcohol as titrant. The HAN–TEAN mixture can be titrated potentiometrically to get distinct endpoints using denatured alcohol with organic additives as the solvent, despite the closely similar pK values of HAN and TEAN. Another method involves the redox iodometric reaction in which the hydroxyl amine is quantitatively oxidised to nitric acid. HAN concentration can be determined by a quantitative oxidation-reduction reaction, where the nitrate salt can be reduced and hydroxyl amine can be either reduced or oxidized depending on experimental conditions (Dotson, 1985). The specific gravity of the aqueous HAN solution can be measured, and the concentration is estimated indirectly using an empirical model (Courthéoux et al., 2006).

$$f = \frac{107.85}{96.042 - w(\mathrm{HAN}) * 30.99},$$ where f=density and $w(\mathrm{HAN})$= wt.% of HAN.

Nitric acid remains as a potential impurity during synthesis and influences the storability and thermal stability of HAN. The total nitrate content in HAN can be estimated by titrimetry or ultra-violet spectroscopy. Titrimetric analysis of nitric acid involves quantitative conversion of HAN to nitric acid using acetone, and titrating with NaOH. Another major component to be analysed is water, and Karl Fischer reagent or Hydranal reagent is used for the determination of water content in HAN.

FTIR and Raman spectroscopy is utilized for the determination of functional groups in HAN (Amrousse et al., 2012). Raman spectrum in the region of 2700–3000 cm^{-1} corresponds to the N–H stretching modes of the hydroxyl ammonium cation in HAN. The intense band at 2732 cm^{-1} in FTIR is a combination of the rotation and the symmetric deformation band of the hydroxyl ammonium ion. ICP-MS is the technique for the detection of metal ion impurities such as iron, copper, titanium, tin, molybdenum, vanadium etc, with part-per-trillion detection limits (Hansen et al., 1989) and to confirm the absence of these metals in <5 ppm level (Schmidt et al., 1990). Thermogravimetric analysis (TGA) is used to determine the decomposition characteristics of HAN, and coupling TGA with mass spectrometry enables the identification of the decomposition products (Yoo et al., 2024).

1.4 Decomposition Characteristics of HAN

HAN consists of a reducing cation, hydroxyl ammonium and an oxidising anion, nitrate. This internal redox pairing leads to complex decomposition mechanism. The decomposition of HAN can proceed via thermal or catalytic reaction, each governed by different reaction pathways. The rate and decomposition temperature are influenced by solution acidity, water content, heating rate and the presence of impurities, each of which is crucial in determining the overall stability of HAN.

1.4.1 Thermal decomposition

The thermal decomposition of HAN solution has been studied using various experimental techniques such as thermogravimetry and differential scanning calorimetry (TG/DSC) (Courthéoux et al., 2006), batch reactors (Yoo et al., 2024), adiabatic calorimeters (Liu et al., 2009), strand burner test (Vosen, 1990) and fast thermolysis apparatus and evolved gas analysis using mass spectrometry (Esparza et al., 2018) or FTIR (Schoppelrei and Brill, 1997).

Solid HAN presumably decomposes more rapidly than HAN-water solutions because of the absence of a diluent, but the induction time of solid HAN is longer than that of aqueous HAN. The longer induction times of solid HAN at low initial set temperatures might have been attributed to a decrease in the chemical activity of species possibly due to strong ion association. The thermal decomposition of HAN is known to be controlled by the rates of collisions of NH_3OH^+ and NO_3^-. The initiation reaction is possibly delayed when the anion and cation are strongly congregated together by hydrogen bonding.

The HAN-based samples with lower water content exhibited more sensitivity to the mechanical impact. On mechanical impact, the HAN-based samples experienced localized heating of their internal components, resulting in the dissipation of some heat energy (Kang et al., 2025). Water performs a stabilizing effect on the hydroxyl ammonium cation and nitrate ion, improving the thermal stability and bringing low impact sensitivity.

Several investigators identified a low-temperature endotherm, associated with evaporation of water, and a major exotherm corresponding to the decomposition of HAN arises from the internal oxidation-reduction reaction in the thermal decomposition profile of aqueous HAN. In aqueous HAN (Amrousse et al., 2015, 2017), the reported onset decomposition temperatures range from 150 to 220 °C. The gaseous species evolved on thermal decomposition of HAN include NO, N_2O, N_2, and O_2. NO_2 is also often detected in the products. Depending on the concentration and experimental conditions, the product composition of HAN decomposition may vary significantly and a widely-cited reaction is as below:

$$4N_2H_4O_4 \rightarrow 3N_2O + 2HNO_3 + 7H_2O$$

Concentrated HAN possess sufficient energy density to self-sustain combustion once thermally initiated. Courthéoux et al., (2002) used a batch reactor to investigate thermal decomposition of binary HAN/water solutions with different

HAN concentrations of 20%, 60%, and 83% by weight. These 20%, 60%, and 83% solutions undergo thermal decomposition at temperatures of 173 °C, 145 °C, and 135 °C, respectively. Decomposition temperature increased with a decrease in concentration of HAN. TG-DTA of 95wt.% HAN (Amrousse et al., 2012) is shown in **Fig. 1.3.** Thermal decomposition studies conducted by Amrousse et al., in a constant volume batch reactor with 95 wt.% HAN showed a pressure increase of 202 mbar corresponding to a decomposition temperature of 152°C.

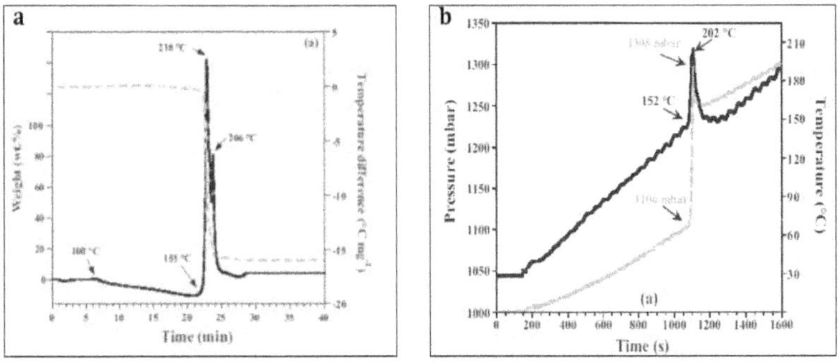

Figure 1.3 (a) TG-DTA of 95 wt.% HAN and (b) Thermal decomposition of 95 wt.% HAN in the batch reactor facility (Amrousse et al., 2012).

Whitmore et al. developed an analytical model to predict the thermal decomposition temperature and decomposition rate for various concentrations of HAN solutions. The model predicts that a 50% HAN solution decomposes thermally and produces a self-sustaining reaction. When the solution concentration is above 60%, energy from decomposition exceeds the latent heat of the water in solution, resulting in a rise in chamber temperature (Whitmore, 2013). The decomposition temperature and energy density as a function of mass fraction of HAN in the solution is presented in **Fig.1.4.**

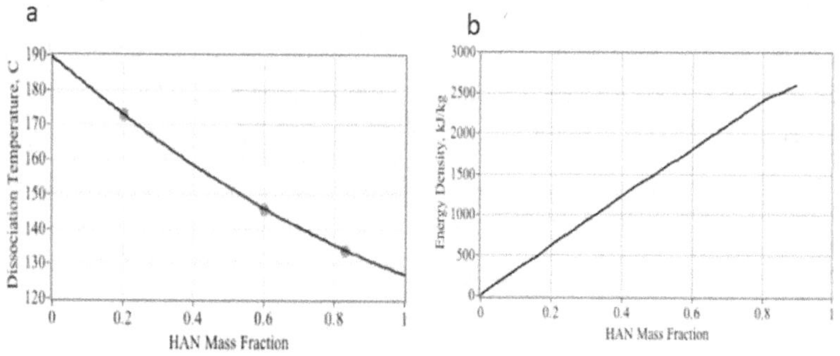

Figure 1.4 (a) Decomposition temperature and (b) energy density as a function of mass fraction of HAN in the solution (Whitmore, 2013).

The decomposition of HAN as proposed by Oxley and Brower (Oxley et al., 1988) commences with a proton transfer from $[NH_3OH^+][NO_3^-]$ forming NH_2OH and HNO_3 (Eq. (1.1)). The nitric acid formed reacts with hydroxyl amine leading to the final products. HAN decomposition is expressed in Eqs. (1.1) to (1.6), wherein Eq. (1.2) is the rate-limiting step.

$$NH_3OH^+ \, NO_3^- \leftrightarrow NH_2OH+HNO_3 \qquad (1.1)$$

$$NH_2OH+HNO_3 \rightarrow HNO+HNO_2+H_2O \qquad (1.2)$$

$$NH_2OH+HNO_2 \rightarrow N_2O+2H_2O \qquad (1.3)$$

$$NH_2OH+HNO \rightarrow N_2+2H_2O \qquad (1.4)$$

$$HNO+HNO_3 \rightarrow 2HNO_2 \qquad (1.5)$$

$$HNO_2+HNO_3 \rightarrow 2NO_2+H_2O \qquad (1.6)$$

N_2O the most prevalent product along with N_2 and NO_2 are identified in the decomposition reaction. N_2O and N_2 are formed by the immediate reaction of hydroxyl amine with the intermediates HONO and HNO (Eqs. (1.3)-(1.4)). Later, Lee and Litzinger (2003) proposed two additional reactions for the formation of N_2O and NO (Eqs. (1.7)-(1.8)).

$$2HNO \rightarrow N_2O+ H_2O \qquad (1.7)$$

$$3HNO_2 \rightarrow 2NO+HNO_3+H_2O \qquad (1.8)$$

Based on the products observed for the decomposition of 80% HAN at 200°C in a fixed bed reactor with a mass spectrometer, Amariei et al., (2007) proposed the following reactions:

$$6N_2H_4O_4 \rightarrow 2N_2 + 2NH_4NO_3 + 6H_2O + 4HNO \quad (1.9)$$

$$2NH_4NO_3 \rightarrow 2NO_2 + 2N_2O + 4H_2O \quad (1.10)$$

N_2 and NO are the primary products in the early stage, whereas N_2O and NO_2 evolve afterwards.

The kinetic parameters for decomposition of HAN as deduced by Lee and Litzinger have $E_a = 63.2 \pm 2.1$ kJ/mol for the proton transfer reaction (Eq. (1.1)). Thermal decomposition of vacuum dried HAN studied using thermogravimetric analysis (TGA), DSC, and mass spectrometry by Esparza et al. (2018). The apparent activation energy for thermal decomposition using TGA and DSC is 62.2 ± 3.7 kJ/mol and 57.5 ± 3.5 kJ/mol, respectively, as calculated by the Ozawa-Flynn-Wall method. The major evolved gases identified by mass spectroscopy are H_2O, NO, N_2O, NO_2 and HNO_3.

Schoppelrei et al., (1996, 1997) observed that the apparent activation energy (E_a) increases as the concentration of hydroxylammonium nitrate (HAN) decreases. Specifically, E_a values range from 60–70 kJ/mol at higher HAN concentrations, rising to approximately 100 kJ/mol at lower concentrations. They proposed that at low concentrations, the overall reaction rate is governed by the oxidation of hydroxylamine by nitric acid (Eq. (1.2)). In contrast, at higher concentrations, the reaction is accelerated by autocatalysis due to the formation of nitrous acid (HNO_2), which enhances heat release and results in a lower apparent activation energy.

Rafeev and Rubtsov (1993) investigated decomposition kinetics of melted hydroxylammonium nitrate. The initial rate of decomposition depends on the concentration of nitric acid formed from HAN dissociation with an activation energy of 15.3±1.8 kcal/mol. They suggested that initial stage proceeds via the interaction of N_2O_5 and $NH_3OH^+$and further accelerated due to the oxidation of NH_3OH^+.

Linear burning rate of aqueous solution HAN increases with an increase in HAN concentration up to 80 wt. % (Katsumi et al., 2010). The effect of pressure on burning rates of HAN-water solutions (50%-95%) has been studied in the pressure range 1-10 MPa. The 80% HAN solution exhibits the highest burning rate compared to 95% solution at a given pressure, but further decrease in concentration decreases the burning rate. The burning rate of HAN may increase or decrease with the pressure effect (McCown et al., 2015). The accelerating effect can be explained by the facilitation of product evaporation and reactive interactions during the decomposition process, and the eventual decline can be attributed to progressive reduction of the liquid-gas interfacial area as it approaches the critical point.

Amrousse et al., (2018) studied the influence of heating rate and reaction temperature on the thermal decomposition of hydroxyl ammonium nitrate. Decomposition of HAN is carried out at 128 °C/min and 8 °C/min. The increased intensity of evolved gases in the case of higher heating rate with similar gas species in both cases is observed. Liu et al made Density functional theory (DFT) calculations using B3LYP/6-311++G(d,p) method and formulated seven different paths for the mechanism of thermal decomposition of HAN (Liu et al., 2017). Influence of heating rate on thermal decomposition of HAN is shown in **Fig. 1.5**. The activation energy of the thermal decomposition calculated by the Kissinger and Ozawa methods is 67.892 and 70.412 kJ/ mol, respectively.

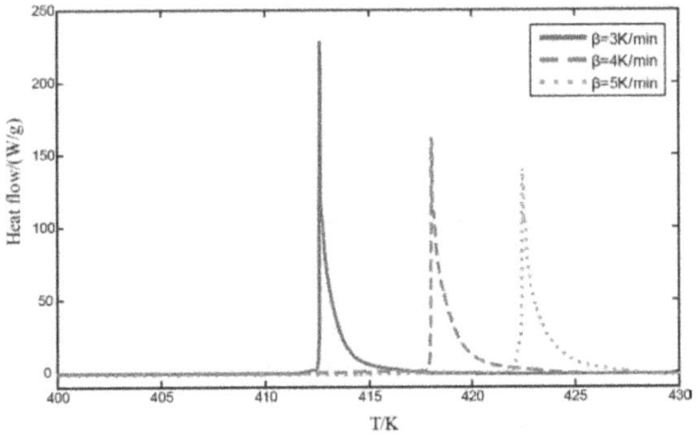

Figure 1.5. DSC curve of HAN at various heating rates (Liu et al., 2017).

Theoretical investigation of thermal decomposition mechanism has been developed by Zhang and Thynell (2018) using the SMD-ωB97X-D method. Multiple kinetic processes are reported in the decomposition, including nitration and nitrosation of hydroxylamine, HNO dimerization, and HONO-regeneration pathways. The autocatalytic pathway of HONO scavenging by nitrosation reactions and regeneration via H-abstraction plays a significant role in the decomposition of concentrated HAN solutions. Sensitivity analysis suggests autocatalysis should be caused by the rise of solution acidity via an acid-catalyzed nitration-nitrosation pathway. HONO is generated autocatalytically in concentrated nitric acid solutions when a small amount of hydroxylamine is present (Eq. (1.11)). Pembridge and Stedman (1979) reported the activation energy of 65 kJ/mol for HONO scavenging, and E_a of 105 kJ/mol for HONO generation.

$$NH_2OH + 2HNO_3 \rightarrow 3HONO + H_2O \qquad (1.11)$$

A kinetic model for the decomposition of hydroxylammonium nitrate studied by Izato et al. (2019) proposed the initial reaction of HAN decomposition by the formation of HNO, HONO and water. The subsequent reaction is HONO abstraction by HAN to give N_2O, HNO_3 and water, and HNO giving N_2O and water. The overall reaction is given in Eq. (1.12).

$$N_2H_4O_4 \rightarrow 0.75\ N_2O + 1.75\ H_2O + 0.5\ HNO_3 \qquad (1.12)$$

1.4.2 Catalytic decomposition

Among the various methods available for HAN decomposition, catalytic decomposition is ideal due to its structural simplicity and reliability. A catalyst can be employed to reduce the activation barrier and enhance the overall reaction rate. **Fig. 1.6** provides a schematic of a monopropellant catalytic engine (Batonneau et al., 2014). The catalyst comprises of a high-temperature-resistant support onto which the active phase is deposited. The active phase is generally precious metals such as Pd, Pt, Rh, Ru, Ir, Au, or Ag, uniformly dispersed across the support surface. Previous studies have demonstrated that noble metals exhibit superior catalytic performance for the decomposition of HAN-based monopropellants, with Ir-based catalysts identified as particularly effective (Wood et al., 1973). The monopropellants are typically ignited through a preheated catalytic fixed bed reactor integrated into a thruster. The injection of HAN on the heated catalyst bed through a remotely controlled valve initiates a highly exothermic decomposition, producing hot gases with temperatures exceeding 1400 °C.

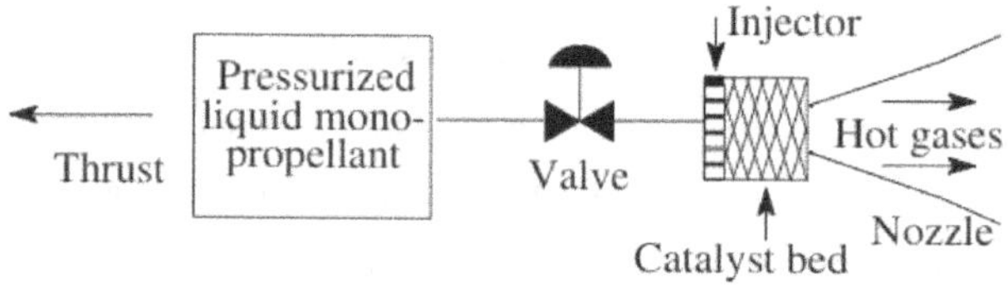

Figure 1.6 Schematic of a monopropellant catalytic engine (Batonneau et al., 2014).

The combustion behaviour of ionic-liquid monopropellants is strongly influenced by the catalyst material, operating pressure, and preheating temperature. A key challenge in catalyst design is to identify materials capable of: (i) initiating monopropellant decomposition at low temperatures, below the thermal decomposition threshold (ii) minimizing the formation of nitrogen oxides during decomposition (iii) maintaining high thermal stability and sufficient specific surface area under extreme temperatures and (iv) preventing loss or sintering of the active phase under propellant flow. A promising strategy to address these challenges is the development of catalysts incorporating a platinum-group metal active phase supported on thermally robust substrate materials. This combination offers enhanced stability, reduced sintering, and improved catalytic efficiency under the demanding conditions of space propulsion applications.

The rate of catalytic decomposition is one order of magnitude higher than that of thermal decomposition (Courthéoux, 2002) and the catalyst can start the decomposition in the presence of water, even for dilute HAN solutions. Courthéoux et al., (2004) demonstrated that aqueous HAN solutions can be catalytically decomposed at remarkably low temperatures (<40 °C) with very short ignition delays (<1 s) using Pt supported on silica-doped alumina xerogels and aerogels. Many catalyst supports are prepared by sol gel method. The method used for introducing the active metal plays a critical role in determining catalytic performance. Catalysts prepared by impregnation exhibited higher activity, which is attributed to the smaller crystallite size of the supported metallic particles. Ren et al. (2007) evaluated a series of alumina- and silica-supported precious-metal catalysts for the catalytic decomposition of hydroxylammonium nitrate using thermal analysis. Among the tested materials, Ir/SiO_2 exhibited the highest catalytic activity, enabling complete decomposition of 80% HAN at 20.7 °C. The results indicate that the choice of support significantly influences the ignition temperature of Ir-based catalysts, highlighting the crucial role of support and metal interactions in determining decomposition performance.

In addition to alumina and silica, numerous other refractory materials have also been explored as catalyst supports. Precious-metal catalysts supported on doped alumina systems, such as Ir/ Al_2O_3–La_2O_3, and CuO/ Al_2O_3–La_2O_3, have been reported for the catalytic decomposition of HAN and ADN monopropellants (Amrousse et al., 2012). The active metal phase is introduced onto the high-surface-area support by an impregnation technique. Amrousse et al. (2012) detailed that the two-stage thermal decomposition of HAN become single stage in the presence of catalyst as given in **Fig. 1.7.** The catalytic decomposition of monopropellants using 10%Ir/Al_2O_3–La_2O_3 proceeds through a single-step mechanism and significantly lowers the decomposition temperature to 111°C.

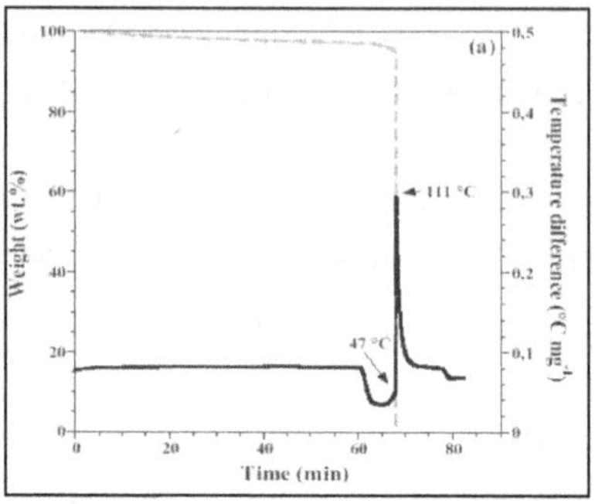

Figure 1.7 TGA showing catalytic decomposition of 95% HAN (Amrousse, 2012).

Ir, Pt, Cu, and Rh supported on hexaaluminates are reported as effective catalysts for the decomposition of ADN and HAN-based liquid monopropellants. Hexaaluminates are hexagonal aluminate compounds with a unique structure and very high temperature stability up to 1600°C due to the peculiar layered crystal structure. Hexaaluminate has inherent properties that make it extremely resistant to sintering. The influence of synthesis procedure and heat treatment on the catalytic activity of hexaaluminate-based supports is studied (Hong et al., 2019). The hexaluminate synthesized via co-precipitation showed high activity towards HAN decomposition due to higher surface areas compared to the same synthesized by sol-gel method.

The catalytic decomposition mechanism using thermal analysis techniques in conjunction with gas chromatography and mass spectroscopy (GC-MS) has been made by Katsumi et al (2015). The study demonstrated that Ir-based catalysts effectively reduce the onset temperature for HAN decomposition, while the product gas distribution remains largely similar to that observed in thermal decomposition. This indicates that the fundamental reaction mechanism is essentially unchanged by the presence of the catalyst. However, catalytic decomposition promotes the formation of nitric acid (HNO_3) at a lower temperature compared to thermal decomposition. The Ir catalyst does not alter the overall reaction mechanism, but accelerates the reaction pathway associated with HNO_3 generation.

The catalytic decomposition of HAN on iridium targets is investigated by Chambreau et al., (2017) and the reaction products are identified by their mass-to-charge ratios and their ionisation energies. The decomposition products include NH_3, H_2O, NO, hydroxylamine (HA), HNO_3, and a small amount of NO_2 at high temperature. The catalytic decomposition of HAN on an iridium surface significantly enhances the production of nitric oxide (NO), as evidenced by the marked increase in the NO/HA ratio. The thermal decomposition mechanism of HAN predicts significant evolution of N_2O, but under the tested catalytic conditions, N_2O is not detected. These findings strongly suggest that, in the presence of the Ir catalyst, a dominant pathway of NO formation proceeds via the decomposition of HA.

Catalysts for HAN decomposition can be configured in various forms, including supported metal pellets, monolithic structures, and metal meshes. HAN exhibits significantly higher viscosity than hydrazine and produces higher combustion chamber temperatures. It tends to disperse poorly upon injection and shows limited wetting on catalyst surfaces (Schmidt, 2023). Consequently, the geometry of the catalyst pellets becomes critical, as it directly influences the available wetted surface area and, the overall catalytic efficiency (Kapteijn et al., 1997).

Amariei et al. (2007) examined the thermal and catalytic decomposition behaviours of 80 wt.% HAN solution. The primary gaseous products identified during decomposition are N_2 and NO. Trace amounts of NO_2 are also detected as secondary species. The influence of catalyst shape on performance is evaluated for powder and spherical shaped catalysts. The results showed that powdered catalysts enabled complete decomposition of the HAN solution. This enhanced activity is attributed to the proper contact between the monopropellant and the catalyst bed, which promotes efficient interaction with active sites. In contrast, sphere-shaped catalysts exhibited lower decomposition efficiency. The reduced performance can be due to preferential paths through the catalyst bed which avoid direct contact between the major part of the propellant and the active centre. The study highlights the critical role of catalyst geometry in governing mass transport, surface accessibility, and efficiency of catalytic decomposition (Hairch et al., 2024).

Amrousse et al. (2016) investigated two distinct catalyst configurations for HAN decomposition: (i) a commercially available granular catalyst, Shell 405 (36% Ir/Al_2O_3, grain size ≈ 25 μm), and (ii) monolithic honeycomb catalysts, as shown in **Fig. 1.8.** During high-temperature combustion testing, the Shell 405 catalyst exhibited significant limitations, including a substantial pressure drop across the catalyst bed and progressive deactivation. The structured design of the monolith minimizes flow resistance, thereby markedly reducing the pressure drop compared to granular catalysts. Furthermore, honeycombs with smaller channels (monolith 600) provide an extended residence time for the decomposed gaseous products within the catalyst channels and hence undergo fast reaction with the solid catalyst surface along with reduced pressure drop. That is, monolithic honeycomb catalysts are advantageous over granular formulations by improving flow dynamics, reducing thermal and mechanical stress, and higher catalytic activity.

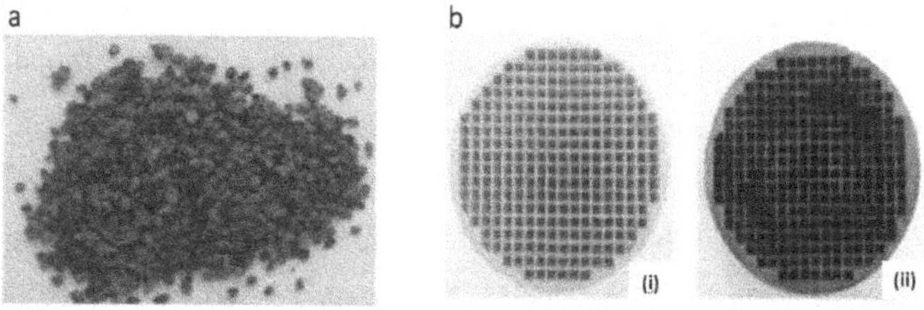

Figure 1.8 Photograph of (a) S405 catalyst and (b) Honeycomb catalyst (i) fresh cordierite and (ii) after deposition of Ir (Amrousse et al., 2016, 2017).

Amrousse et al. (2017) investigated the catalytic decomposition of the HAN-based monopropellant SHP163 using several catalyst systems, including the granular S405 catalyst, Ir-coated honeycomb monoliths, and Ir-CuO honeycomb structures. Their results showed that SHP163 undergoes a highly exothermic, single-step decomposition with a remarkably low onset temperature of approximately 65 °C when exposed to these catalysts. The TGA showing the thermal and catalytic decomposition of SHP163 is shown in **Fig. 1.9.** Thruster firing tests further revealed notable differences in performance among the catalyst types. Honeycomb catalysts, owing to their structured geometry and improved gas flow characteristics, produced significantly steeper pressure rise slopes, indicating faster and more efficient combustion. Ir-CuO honeycomb catalyst demonstrated superior burning behaviour and improved overall thruster response. Amrousse et al. (2015) showed that a 20 wt.%Ir-10wt.%CuO biphasic catalyst achieved pressure slopes comparable to those of pure Ir honeycomb catalysts, while offering substantial economic advantages due to reduced noble-metal content. The pressure gradients are steeper with Ir honeycomb catalysts at 400 and 600 cpsi (cells per square inch) in comparison to the S405 catalyst. **Fig. 1.10** shows the pressure profile as a function of time during HAN burning test using different catalysts. Ir-CuO honeycomb catalyst accelerates the second pressure slope, and the maximum pressure value (0.4 MPa) has been reached in a short time (Amrousse, 2021).

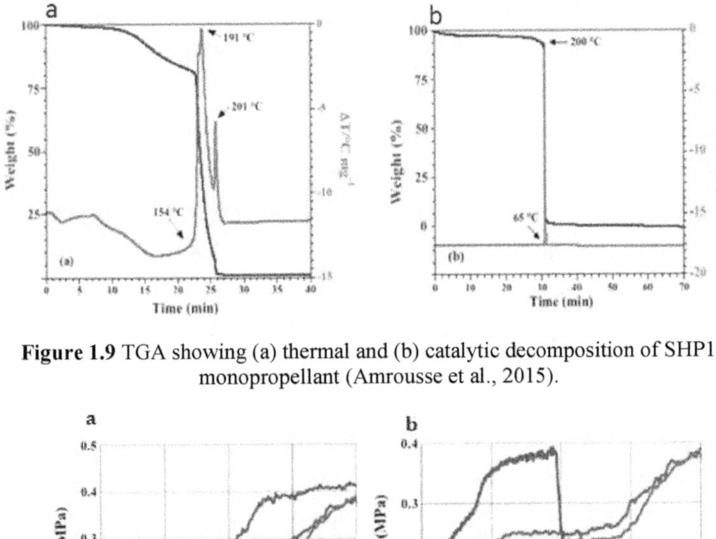

Figure 1.9 TGA showing (a) thermal and (b) catalytic decomposition of SHP163 monopropellant (Amrousse et al., 2015).

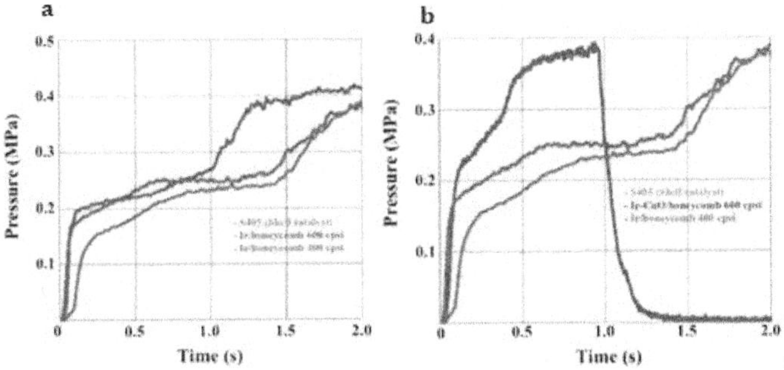

Figure 1.10 Pressure profile as a function of time during HAN burning test showing the effect of catalyst and its cell size (a) Ir honeycomb catalyst of two cell sizes compared to S405 catalyst and (b) Ir–CuO honeycomb and Ir honeycomb catalysts compared to S405 catalyst (Amrousse et al., 2017).

Zhou et al., (2023) investigated the catalytic decomposition of HAN on the Ir (110) surface using density functional theory (DFT). The study identified four possible decomposition pathways for the HAN molecule adsorbed on Ir(110). In most cases, the reaction is initiated by the scission of the N–O bond within the NH_2OH moiety. However, cleavage of the N–O bond in the NO_3^- group also plays a crucial role in determining the overall catalytic decomposition pathway.

Several classes of materials such as simple oxides, mixed oxides, and perovskite have been investigated as catalyst supports for HAN decomposition. Apart from optimizing the catalyst geometry, the cost of precious active metals remains a significant limitation in catalyst design. Consequently, considerable research efforts have been directed toward substituting these noble-metal catalysts with more economical metal oxides. Catalysts known to be active for N_2O decomposition are of particular interest, as N_2O is a key intermediate in the HAN decomposition. The catalytic activity for HAN decomposition in 13.5–17.8 wt.% Mn, Co, Cu, Pt, or Ir metallic oxides impregnated on the cordierite (MgO-SiO_2-Al_2O_3) honeycomb support are thermally evaluated (Yoo et al., 2024). The study showed that the pore structure of the metal oxide-honeycomb catalysts is largely independent of the type of metal incorporated. Among the tested formulations, the Cu/honeycomb catalyst exhibited the highest catalytic activity.

Cobalt-doped cerium oxide (Agnihotri et al., 2018) effectively lowers the HAN decomposition temperature and increases the reaction exothermicity. A catalyst containing an optimal cobalt loading demonstrates stable catalytic activity in batch-reactor tests even after 50 pulse injections. This performance is primarily attributed to the Ce^{3+} species, which enhances the Ce^{4+}/Ce^{3+} redox cycling, improves the reducibility of ceria, and promotes the formation of oxygen vacancies. However, excessive cobalt loading leads to a decline in catalytic performance. Among the tested compositions, Ceria with 26% Co loading (CeCo26) shows the highest activity toward HAN decomposition. The low and high magnification SEM images of different catalysts and DSC plot of HAN with the catalysts are shown in **Fig.1.11.** The same research group further examined the advantages of a bifunctional catalyst combining Ir as the active metal with the Co doped ceria support (Agnihotri and Oommen, 2020). In batch-reactor experiments, the Ir/CeCo26 catalyst produced a high-pressure build-up and maintained structural integrity for approximately 45–50 pulse injections, demonstrating both enhanced activity and durability.

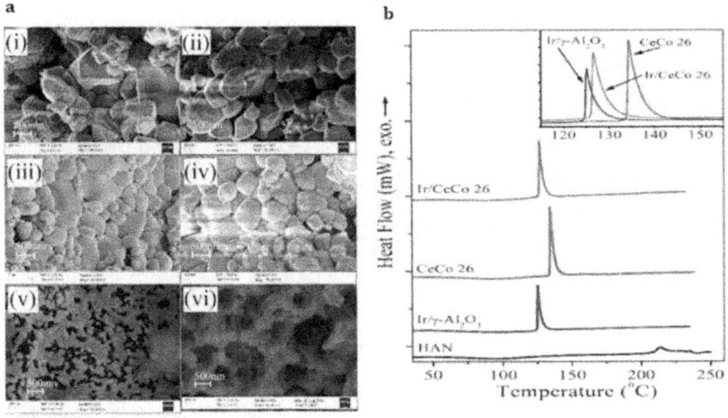

Figure 1.11 (a) Low and high magnification SEM images of pellet form for different catalysts (i & ii- CeCo 26, iii & iv- Ir/CeCo 26, v & vi- Ir/γ-Al_2O_3) and (b) DSC of HAN with different catalyst systems(Agnihotri and Oommen, 2020).

Agnihotri and Oommen (2021) compared a newly developed ceria-based catalyst with an iridium-coated alumina catalyst for the decomposition of the ternary monopropellant HAN: Methanol: Water (70:15:15 w/w). In contrast to binary HAN formulations, the ternary blend produces much higher temperatures and pressures, providing a more realistic environment for evaluating catalyst durability under thrusters like conditions. A batch reactor is employed to investigate both catalytic and non-catalytic decomposition for assessing long-term catalyst performance and deactivation. The results show that the ceria-based catalyst maintains its structural and chemical integrity better than the Ir-based catalyst under these severe conditions. The findings demonstrate the suitability of ceria-based catalysts for HAN-based monopropellant applications.

Zhou et al. (2025) reported the development of low-cost catalyst for HAN decomposition using a Ru-Pt alloy nanoparticle impregnated on mesoporous alumina support. Ruthenium having high catalytic activity and a relatively low price are prone to oxidation. Alloying with Pt forming Ru-Pt nanoparticle catalyst exhibited a higher activity and stability than the Ru nanoparticle catalyst alone. The catalytic decomposition mechanism of HAN on the Ru(101) surface using Density Functional Theory (DFT) calculations is investigated (Shen et al., 2026). The catalytic decomposition of HAN on a Ru surface begins with the decomposition of the nitrate group with the cleavage of N-O bond generating NO_2 and O, followed by the decomposition of the hydroxylamine group where O reacts with NH_3OH, abstracting two H atoms to form H_2O and NH_2O. However, on the Ir surface, the catalytic decomposition begins with NH_2OH cleavage. HAN has a lower energy barrier and higher adsorption energy on the Ir surface, which is beneficial to further decomposition reactions.

1.5 HAN-based propellant formulations

Hydroxyl ammonium nitrate acts both as an oxygen-rich oxidizer and a fuel component, yielding a net positive oxygen balance. To approach stoichiometry and enhance performance, HAN-based monopropellants are typically blended with additional fuels. Appropriately formulated HAN–fuel monopropellants can achieve improved specific impulse values approaching those of bipropellant systems while generating high adiabatic combustion temperatures. The polar nature of HAN aids in the solubility in water or other solvents to form liquid propellants (Hoyani et al., 2017). Varying the type and proportion of added fuel, as well as the water content, provides a convenient means to tune the formulation's energy content, adiabatic flame temperature, and physicochemical properties. HAN-based formulations are non-carcinogenic, non-flammable, and non-explosive at atmospheric pressure, and have a vapor head composed only of water. Among the various green monopropellant formulations, HAN is most promising due to its high specific impulse, reduced freezing point, high density, and density-specific impulse. **Table 1.7** displays the performance characteristics of different green monopropellants.

The fuel (or reducer) can be ionic or molecular compounds. One of the advantages of ionic fuels over molecular species is to increase the oxidizer solubility. Selection of a fuel depends on its heat of formation, density, compatibility, stability, solubility, low toxicity and volatility (Wucherer et al., 2000). Different candidate fuels used in HAN formulations are, triethanolammonium nitrate (TEAN), glycine, hydroxyethylhydrazinium nitrate (HEHN), methanol etc. **Table 1.8** gives the composition of various HAN-based propellant formulations.

The development of HAN-based formulations began with liquid gun propellant (LGPs) using amines as fuel. The LGP formulations, LP1846 and LP1845, used tri-ethanol-ammonium nitrate (TEAN) and LP1898 used diethyl hydroxyl ammonium nitrate (DEHAN) as fuel (Decker et al., 1987, 1994). LP1846 can achieve complete combustion when the HAN: TEAN mol ratio is 7:1(Eq. (1.13_). Despite the stoichiometric ratio, soot deposition is observed in rocket propellant studies, which can be eliminated by using an excess of oxidiser or replacing it with a fuel containing less number of carbon atoms (Schmidt, 2023). The ternary amine based formulation exhibits a stabilising effect towards decomposition reactions of the HAN, but was inappropriate for rocketry applications due to the high combustion temperature (2230 °C)(Kasbi et al., 2025).

$$([OHCH_2CH_2)_3N^+H]\ [NO_3]^- + 7N_2H_4O_4 \rightarrow 6CO_2 + 22H_2O + 8N_2 \qquad (1.13)$$

HANGLY26 is another HAN-fuel formulation consisting of 60% HAN, 14% glycine, and 26% water by weight. The burning rate of HANGLY26 is two orders of magnitude lower than LP based formulation, though both have the same amount of HAN and water. This is attributed to the difference in burning characteristics of the fuel component present in the propellant (Chang et al., 2001). HANGLY26 has a lower Isp than hydrazine (190 seconds) but is significantly denser than hydrazine and produces greater volumetric impulse efficiency. Methanol based formulation, such as HAN269MEO perform better than glycine formulations, achieving a vacuum Isp near 270 s with higher adiabatic temperatures limiting the lifetime of catalysts (Whitmore et al., 2013). The stoichiometric reaction of HAN and methanol is shown in Eq. (1.14).

$$2N_2H_4O_4 + 2CH_3OH \rightarrow 2CO_2 + 10H_2O + 3N_2 \qquad (1.14)$$

Another HAN-based green propellant that has been developed at the Institute of Space and Astronautical Science (ISAS), Japan Aerospace Exploration Agency (JAXA), is SHP163 (Amrousse et al., 2017) which contains 73.6 weight % HAN, 3.9 weight % ammonium nitrate (AN), 16.3 weight % methanol, and 6.2 weight % water. Ammonium nitrate (AN) lowers the melting point of the solution and reduces the water content. Methanol is selected as fuel because of its favourable combustion characteristics and improved specific impulse. SHP163 has a density of 1.4 g /cm^3 and volumetric specific impulse of 396 g s /cm^3, with flame temperature rises to about 2130 °C. The theoretical Isp of SHP163 is 254 s, which is 20% higher than that of hydrazine when the combustion chamber pressure is 0.7 MPa at 50:1 nozzle expansion ratio. Katsumi and Hori examined the combustion of SHP163 and identified hydrodynamic instability as the principal trigger for the observed high burning rates (Katsumi et al., 2021). This is governed by the formation rate of water nuclei, which is induced by superheating from the exothermic HAN decomposition. Hori et al., (2019) reported on the testing of SHP 163 in space as part of the Rapid Innovative Payload Demonstration Satellite (RAPIS) project. The 1 N Green Propellant Reaction Control System (GPRCS) engine powered by SHP163 has been integrated into RAPIS-1, which began operating in the space environment in January 2019. The onset of SHP163 is observed at 154°C with two exothermic peaks at 191 and 201°C **(Fig. 1.9)**. The reaction pathway of HAN-Methanol system by quantum chemical method identified the mechanism of initiation as oxidation of methanol, $CH_3OH + OH^{\cdot} \rightarrow {}^{\cdot}CH_2OH + H_2O$, followed by further reactions (Schmidt, 2023).

ASCENT (Advanced Spacecraft Energetic Non Toxic) propellant, formerly known as AF- M315E, is an advanced monopropellant formulation developed by the Air Force Research Laboratory (AFRL)(Masse et al., 2012). AF-M315E has been flown on the GPIM (Green Propellant Infusion Mission) with thrusters developed by Aerojet Rocketdyne. It consists of a mixture of HAN, HEHN (hydroxyethyl-hydrazinium nitrate) and water (Giordano, 2023). AF-M315 E has a 13% increase in specific impulse and a 63% increase in density, making it more effective for miniaturising propulsion systems. The adiabatic flame temperature of

AF-M315E is much higher than that of hydrazine (~1830 °C) which makes it difficult to rapidly produce an economical and simpler thruster design.

The High-performance Non-detonating Propellant (HNP) family is developed by IHI Aerospace Co., Japan. HNP209, HNP221, and HNP225 are members of this green monopropellant family and are made of HAN, hydrazinium nitrate (HN), methanol, and water (Fukuchi, 2008). HAN/HN-based monopropellants had a greater density, ranging from 1.4 to 1.5 g/cm^3, exhibits superior density specific impulse compared to their hydrazine-based counterparts (Igarashi et al., 2019). Their low adiabatic flame temperature, when compared to other energetic ionic liquid monopropellants such as AF-M315E and SHP163 makes them unique. HNP225 is the one among the family with the least adiabatic flame temperature, approximately 727 °C, even less than hydrazine (~927 °C), and delivers theoretical vacuum specific impulse almost 200 s. On the other hand, HNP221 and HNP209 demonstrate the theoretical specific impulses of 241 and 260 s, respectively, when the expansion ratio is 100:1 and the chamber pressure is 1.0 MPa.

The HNP family of green monopropellants ignite on catalytic decomposition. Igarashi et al., (2017) performed tests with newly developed proprietary catalysts and showed excellent response and stability compared to hydrazine, either in pulsed or continuous modes of operations, with preheating temperatures starting from 200 °C for HNP221 and HNP225 monopropellants.

For binary HAN propellant system, thermal decomposition starts after complete vaporisation of water. The same phenomenon is observed with HAN/methanol systems also. The studies on the pressure dependence of the burning rate of two HAN/methanol-based liquid monopropellants, HAN269MEO15 and HAN284MEO17, reveal distinct combustion characteristics (Schmidt, 2023). The study theorised that the boiling processes of both water and methanol in the propellant have a significant influence on the combustion behaviour. This interpretation is supported by the presence of temperature plateaus in the temperature profiles measured across the reaction zones of the propellant.

The linear burning rate for binary HAN mixture increases with HAN concentration up to 80% and then decreases (Katsumi et al., 2012). The high burning rate is due to the intense boiling of water owing to the superheat mechanism. Azuma (2013) studied the influence of ammonium nitrate (AN) and methanol concentration in HAN-based propellant formulation. An excess of AN increases the burning rate and decreases the decomposition temperature, which is theorised as AN altering nitric acid/HAN molar ratio and in turn influences reactivity of HAN. As per the studies of Katsumi et al., (2009), the increase in water content increases the burning rate and methanol suppresses bubble growth kinetics which decreases the rate of evaporation and hydrodynamic instabilities. A steady increase in burning rate over the 4–30 MPa pressure range can be achieved by gelling the propellant as reported by Ferguson et al. (2021).

Table 1.7 Comparison of performance characteristics of Green monopropellants.

Parameters	**Hydrazine**	**SHP 163 (HAN-based)**	**AFM 315 E (HAN based)**	**ADN based (LMP 103S)**	**H_2O_2**
Freezing point (°C)	2	-30	-22	-6	-7
Density (g/cc)	1.0	1.4	1.5	1.4	1.3
Theoretical Isp (s)	239	276	266	255	182
Density specific impulse (g cm^3 s)	240	386	399	332	256
Adiabatic flame temperature (K)	1183	2166	2180	2054	1154

Table 1.8 Composition of different HAN-Fuel formulations (Nosseir et al., 2021).

Propellant	Composition (wt.%)								
	HAN	TEAN	DEHAN	HEHN	MeOH	AN	HN	Glycine	H_2O
LP1846	60.8	19.2							20.0
LP1845	63.2	20.0							16.8
LP1898	60.7		19.3						20.0
HANGLY26	60.0							14	26.0
HAN269 MEO15	69.7				14.8	0.6			14.9
HAN284 MEO17	77.3				17.2	0.7			4.9
SHP069	81.9				6.9	4.3			6.9
SHP 163	73.6				16.3	3.9			6.2
HNP115	46.0	25.0					23		6.0
HNP 221									
HNP 225									
AF-M315E	34			34		5			27.0

1.6 Ionic liquid fuels for space applications

Ionic liquids (ILs) are ion pairs usually composed of an organic cation and an organic/inorganic anion possessing unique properties. The choice of cation is usually quaternary ammonium ion or an N-heteroatomic cation. Among various heteroatomic cations such as imidazole, pyrazole, triazole and tetrazole. Imidazolium-based ionic liquids are frequently selected owing to their superior chemical stability and liquid-phase properties relative to analogous triazolium and tetrazolium salts. The common heterocyclic nuclei used for the construction of ionic liquids is presented in **Fig. 1.12.**

The most distinctive feature of ILs lies in their structural designability. To achieve low melting points, ionic liquids employ bulky, asymmetric ions with poor packing efficiency, which effectively reduces their lattice energy, while the incorporation of amino groups tends to diminish sensitivity. Conversely, propargyl or allyl substitution has been reported to increase reactivity and can promote hypergolic ignition in appropriate oxidiser pairings (Liu et al., 2017; Zhang et al., 2011). Ionic liquid fuels effectively address the limitations associated with conventional molecular fuels, such as volatility, toxicity, and handling hazards. They offer several inherent advantages, including high thermal and chemical stability and negligible vapour pressure.

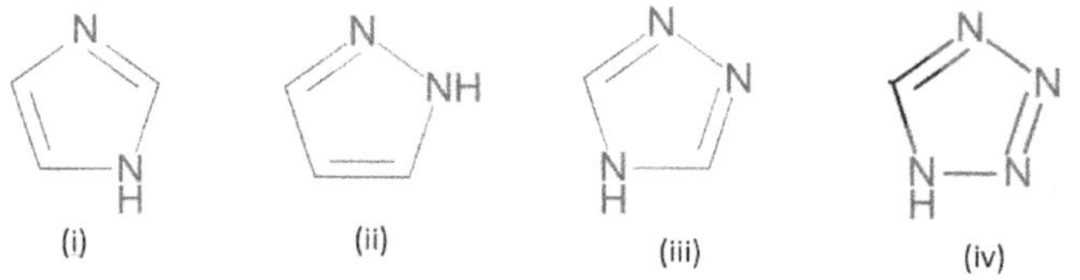

Figure. 1.12 Heterocyclic nuclei used for constructing energetic ionic liquids. (i) imidazole, (ii) pyrazole, (iii) triazole, (iv) tetrazole.

Energetic ionic liquids (EILs) are a new class of low melting energetic materials explored as environmentally benign explosives or propellant fuel (Sebastiao et al., 2014; Zhang and Shreeve, 2013; Zhang et al., 2011). A major innovation in the 21st century is the discovery that ionic liquids function as fuels, which may exhibit hypergolic reactions with suitable oxidisers (Zhang et al., 2011). These low melting salts have since become the key in the field of energetic material research. ILs combine the useful characteristics of salt systems with the advantage of handling liquid systems. Their low sensitivity on impact, friction, shock, and electrostatic discharge further adds to the credit (Zhang et al., 2014). For propellant applications, the EILs must possess high energy density, which is associated with a positive heat of formation (HOF) (Smiglak et al., 2010). N-heteroatomic ions exhibit energetic characteristics when paired with appropriate anions, which originates from the energetic N-N and C-N bonds (Singh et al., 2006). The high heat of formation contributes to high combustion chamber temperature and eventually high specific impulse (Wang et al., 2012).

A major obstacle in the development of novel propellants is establishing an appropriate oxygen balance between the fuel and oxidizer within a single formulation (Christe and Drake, 2003). The oxygen balance is the mass percentage of oxygen that is theoretically released or consumed on decomposition of the oxidizer or the fuel. Oxidizers present a positive oxygen balance whereas fuels display negative oxygen balance. Most EILs demonstrate lower performance than traditional energetic materials due to insufficient oxygen content to achieve complete oxidation (Jones et al., 2006). Hence, EILs must be combined with suitable oxidisers to fully exploit their energetic potential. Hydroxylammonium nitrate (HAN), which exhibits a positive oxygen balance, is a versatile oxidiser for pairing with EIL fuels.

1.6.1 HAN-Ionic liquid fuel formulation

Most of the studies on HAN formulation are based on blends of HAN with hydrocarbon fuels for chemical propulsion applications (Amrousse et al., 2017; Kasbi et al., 2025). Currently, a new class of HAN propellant with imidazolium-based ionic liquid has been proposed as dual mode propulsion system (Rovey et al., 2020). Multi-Mode Propulsion (MMP) is a class of propulsion system that shares propellant between two or more 'modes' of propulsion. The propellant is held in a single, shared tank and can be allocated between the propulsive modes as needed. A wide range of MMP concepts has been studied, most of which is dual mode combines chemical and electric modes.

In dual mode propulsion, Berg and Rovey (2012) investigated imidazolium-based ionic liquids fuels with hydroxyl ammonium nitrate in a binary monopropellant mixture. In a binary monopropellant mixture with 60-70% HAN oxidizer and ionic liquid components shows 1-4% improved specific impulse, exceeding the current state-of-the-art propellants (Berg and Rovey, 2013; Donius and Rovey, 2011). Ionic liquids with more oxidising elements in the anion require less additional HAN oxidiser to achieve required performance level. The electric and chemical propulsion features of a dual mode mixture of HAN and 1-ethyl-3-

methylimidazolium ethyl sulfate, [EMIM][$EtSO_4$] have been the subject of only preliminary study.

Fonda-Marsland and Ryan (2017) also investigated ILs for chemical-electrospray propellants, and described several selection metrics for estimating the performance of each propellant in both modes. The study concluded that reducing the number and size of hydrocarbon groups and increasing the highly electronegative species, such as oxygen and nitrogen, reduces carbon in the monopropellant exhaust which also increases the surface tension and specific conductivity of the propellant required for electrospray mode.

Berg and Rovey evaluated decomposition of propellant mixtures of HAN with 1-butyl-3-methylimidazolium nitrate ([BMIM][NO_3]) and [EMIM][$EtSO_4$] on a heated catalyst bed (Berg and Rovey, 2012, 2013) in a microreactor. HAN-[BMIM][NO_3] yield a pressure rise to 26 mbar/s on rhenium catalyst preheated at 160°C compared to 14 mbar/s for iridium catalyst. HAN-[EMIM][$EtSO_4$] shows slightly less activity on decomposition with a pressure rise of 20 and 4 mbar/s on rhenium and iridium catalyst, respectively. The study suggests that the binary mixture requires a higher preheat catalyst bed temperature with a supported rhenium catalyst to achieve a comparable performance to hydrazine.

Berg and Rovey's recent work focused on FAM-110A (Froberg Aerospace Multimode Propellant 110A), which is a 41:59 by weight blend of [EMIM][$EtSO_4$] and HAN. They demonstrated that platinum is an effective catalyst for FAM-110A (Mundahl et al., 2016) and the ionic liquid propellant exhibits stable electrospray emission in both cation and anion extraction modes, with the highest achieved specific impulse of 412.37 s (Berg et al., 2015). The propellant blend tests on platinum, rhenium and titanium catalyst bed design carried out by the same team (Berg and Rovey, 2013) showed a higher rate of decomposition at 85°C with low activation energy using Pt catalyst. Pt is suggested as the prime catalyst material for thruster development, but its susceptibility to melting at the high flame temperatures limits its use. Since the mechanism of the reaction for the monopropellant blend is initiated by HAN decomposition and reaction of intermediate species with the fuel, the study proposes that lowering of flame temperature can be accomplished by increasing the fuel concentration in the blend. Pt is a prime choice for catalyst material if thermal management is addressed.

Mundahl et al., (2017) investigated on choline nitrate-glycerol as a fuel component in a binary mixture with oxidizers ammonium nitrate (AN) and HAN. Choline nitrate-glycerol with HAN has predicted specific impulse 10% higher than [EMIM][$EtSO_4$]-HAN mixture but former has a 26–88% higher decomposition temperature. The study concluded that [EMIM][$EtSO_4$]-HAN showed promise as an electrospray propellant with platinum catalyst and demonstrated both robust chemical performance and admirable electric propulsion performance offering versatile capability to address complex and emerging space mission necessities. The inclusion of [EMIM][$EtSO_4$] increase carbon content and molecular weight, thereby reducing the propellant's ignitability. **Table 1.9** gives a comparison of performance parameters for HAN and HAN-ionic liquid formulations.

Table 1.9 Performance parameter of HAN and HAN-ionic liquid fuel formulations.

Propellant	Fuel	Isp (s)	Flame temperature (°C)
HAN	-	235	~1800
HAN-[BMIM][NO_3]	1-butyl-3-methyl-imidazolium nitrate	302	~2800
HAN-[EMIM][$EtSO_4$]	1-ethyl-3-methyl-imidazolium ethyl sulphate	298	~2500

1.7 Motivation and Scope

HAN-based propellant formulations, such as SHP163 and AF-M315E offer superior performance compared with hydrazine, but their practical implementation is still limited by several challenges such as high adiabatic flame temperature, the requirement of robust preheated catalyst beds, material compatibility issues, thermal instability, and ignition delays. An understanding of the interaction of HAN with different fuels and its influence on the combustion is required to develop an optimised HAN fuel formulation. The thesis aims to investigate the role of fuel, especially imidazolium-based ionic liquid fuels, in enhancing the energetics and combustion performance of HAN-based monopropellants. In addition, the development of robust catalysts is critical for the real-world application of HAN propellants. Conventional Ir/Al_2O_3 catalysts cannot withstand the high adiabatic flame temperatures generated during HAN decomposition. Therefore, this work also focuses on developing thermally stable perovskite-based catalyst materials with enhanced catalytic activity and durability for HAN decomposition. Hence, the thesis seeks to address the existing gap in efficient catalyst systems and optimised ionic liquid-based fuel formulations for advanced HAN-based green propulsion systems.

1.7.1 Potentiality of HAN-fuel formulation

HAN-based propellant formulations comprise hydroxylammonium nitrate (HAN), water, and fuel component. Two widely referenced high-performance formulations are SHP-163 (74% HAN, 4% ammonium nitrate (AN), 6% H_2O, and 16% methanol) and AF-M315E (a mixture containing HAN, HEHN, AN and water), developed by JAXA and the U.S. Air Force Research Laboratory (AFRL), respectively. Most studies on HAN have focused on aqueous HAN solutions or HAN-methanol blends. Methanol has been a preferred fuel because its lower number of carbon atoms reduces the likelihood of soot formation during decomposition. Additionally, HAN-methanol mixtures have been reported to exhibit lower decomposition temperatures.

The fuel component plays a critical role in determining the energetic performance, thermal stability, and decomposition behaviour of the HAN-based propellant. Tailoring the fuel chemistry provides opportunities to design HAN formulations optimized for specific propulsion requirements. Identifying the most suitable HAN-fuel combination remains an active research area. Continuous systematic investigation is essential to develop the next generation of HAN-based green propellants.

HAN-EIL propellant formulation possesses high solubility, density and negligible vapor pressure. The benign characteristics along with improved performance parameters make EIL an ideal choice for monopropellant applications. The advantage of using N-rich EILs in propellant formulation is that the high energy that can be harvested due to their positive heat of formation. Also, the major decomposition product is N_2 which eliminates carbon deposition on incomplete combustion and improves the overall performance of monopropellant decomposition.

1.7.2 Novel Catalyst for HAN decomposition

The decomposition of HAN can proceed via thermal or catalytic reaction. The catalyst can be employed for reducing the activation barrier and enhancing the overall reaction rate. Typical catalyst system comprises a high-temperature-resistant support onto which the catalytically active phase is deposited. However, HAN decomposition is highly exothermic, releasing substantial heat with an adiabatic flame temperature exceeding 1800 °C. Such extreme temperatures pose severe challenges for catalyst durability. Conventional catalysts such as Shell 405 and Ir-doped hexaaluminate are unable to withstand these thermal loads, leading to structural degradation, active-phase sintering, and loss of catalytic activity.

The challenge in the development of next-generation catalysts is to identify materials capable of initiating monopropellant decomposition at low temperatures, maintaining high thermal stability and preventing loss or sintering of the active phase. A promising strategy to address these challenges is the development of catalysts incorporating a platinum-group metal active phase supported on thermally robust substrate materials to overcome the high reaction temperature.

1.8 Objective of the work

HAN-fuel monopropellants can achieve improved specific impulse values approaching those of bipropellant systems while generating high adiabatic combustion temperatures. Selection of a fuel depends on its heat of formation, density, compatibility, stability, solubility, low toxicity and volatility. Fuels may be molecular (e.g., alcohols) or ionic-liquid-based. The ionic liquid fuels offer several advantages such as low vapor pressure, higher density, lower freezing point, and often enhanced storability and handling safety.

Ionic liquid's unique advantage is selecting a proper cation and anion to obtain a compound with desired properties, which directly translates into potential applications in various fields. Energetic ionic liquids (EILs) are a special class of ILs that have attracted much attention as environmentally benign explosives or propellant fuels due to good thermal stability, excellent energetic performances, low volatility, wide liquid range, and reduced sensitivity. Consequently, there has been growing interest in exploring ionic liquid fuels in HAN formulations to achieve improved performance.

Iridium-based catalyst systems have been widely employed for monopropellant decomposition since the development of the conventional Ir/Al_2O_3 catalyst. Despite their effectiveness, the use of noble metals like iridium presents major limitations due to their high cost. Accordingly, research efforts are directed towards developing catalysts with reduced iridium loading or exploring low-cost, non-noble metal substitutes that can deliver comparable catalytic activity and stability.

Current work is divided into two parts, (i) studies on ionic liquid fuels for HAN formulations and (ii) decomposition of HAN using a novel catalyst. The specific objectives of the work include:

1. Explore the suitability of HAN-Fuel systems for propulsive applications through theoretical studies
2. Theoretical performance evaluation of selected ionic liquid (IL) fuels and HAN using NASA-CEA Computer code
3. Synthesis of selected IL fuels and their characterization
4. Investigating the decomposition mechanism and kinetics of IL Fuel and HAN-fuel system for designing a suitable monopropellant formulation

5. Development of suitable catalyst systems and evaluation of HAN monopropellant decomposition using the catalysts.

1.9 Thesis Organisation

Chapter 1 provides a background to chemical propulsion systems and highlights the need for green monopropellants as safer alternatives to hydrazine. It introduces hydroxylammonium nitrate (HAN) and energetic ionic liquids (EILs) as promising candidates. This chapter summarises previous studies on HAN including thermal stability, decomposition behaviour, catalyst development, and fuel formulation, thereby establishing the rationale for the present work.

Chapter 2 details the synthesis of HAN via neutralization, its concentration to the range required for propellant application and characterisation using chemical and instrumental methods. The thermal properties of HAN are evaluated by TG and DSC analysis. Batch reactor testing of HAN is carried out to determine the pressure rise and ignition delay time. The theoretical performance ($Isp_{(vac)}$) evaluation of 80% and 95% HAN using the NASA-CEA computer code is also discussed in comparison with that of hydrazine.

Chapter 3 investigates the application of ionic liquid fuels in HAN formulations. Density Functional Theory (DFT) is used to screen ionic liquids based on 1-butyl-3-methyl imidazolium (BMIM) cation and five energetic anions viz., nitrate(N), dinitramide (DN), 3-amino-1,2,4-triazolate (Tz), 3,5-diamino-1,2,4-triazolate (DTz) and 5-amino tetrazolate (AT). The computational insights help identify promising energetic fuel candidates. 1-butyl-3-methylimidazolium-5-aminotetrazolate (BMIMAT) is selected as the candidate fuel due to the high heat of formation. The BMIMAT is synthesised via metathesis reaction and characterized. Thermal behaviour of BMIMAT is assessed using DSC/TGA and Py–GC–MS. The mechanism of decomposition proposed based on product analysis by Py–GC–MS is complemented by theoretical studies. HAN-BMIMAT system delivers maximum theoretical Isp of 324 s at an O/F ratio of 6 with a chamber temperature of 2600 °C. Monopropellant formulations are prepared by combining

80% HAN as the oxidizer with BMIMAT energetic fuel. The thermal behaviour of HAN-BMIMAT formulations is also studied and the results are presented.

Chapter 4 investigates the influence of alkyl side chain (ethyl, butyl, allyl, vinyl) of the cation on the energy content, thermal stability, and reactivity of ionic liquids containing energetic anions. DFT results, HOMO–LUMO gaps, heats of formation, and predicted propulsion performance are compared to identify optimal structures for HAN-based fuels. The vinyl-substituted cation paired with the 5-aminotetrazolate (AT) anion (VMIMAT) is highly energetic owing to its high heat of formation; however, its low HOMO–LUMO energy gap (2.8 eV) indicates increased sensitivity and reduced stability. In contrast, the allyl-substituted AMIMAT exhibits a high heat of formation (485 kJ/mol) while offering improved stability, making it a more promising fuel candidate. Consequently, AMIMAT is selected and synthesized for the first time via a metathesis reaction. The HAN–AMIMAT system delivers a vacuum specific impulse (Isp) of approximately 320 s, which is significantly higher than that of 95% HAN. Monopropellant formulations were prepared using 80 wt.% HAN as the oxidizer and AMIMAT as the energetic fuel and the thermal behavior of the formulations was investigated, and the corresponding results are presented.

Chapter 5 aims to develop a low-cost catalyst as a promising alternative to the Pt and Ir based catalyst system for HAN decomposition. The study focuses on the synthesis of catalysts based on lanthanum chromite perovskites substituted with transition metals (Co, Ir) and their structural characterization. The influence of a green capping agent in improving the activity of the catalyst is studied. The catalytic performance evaluation is carried out by thermal and batch reactor test. Co doped lanthanum chromite of composition $LaCr_{0.8}Co_{0.2}O_3$ (LCCO-AV) synthesised using aloe vera gel capping agent exhibit catalytic activity with reduction in onset decomposition temperature of HAN by 35 °C and the peak decomposition temperature by 54 °C. However, Ir doped lanthanum chromite with an optimal composition of $LaCr_{0.9}Co_{0.1}O_3$ (LCIO-AV) exhibits better catalytic activity with a further lowering of the onset temperature of decomposition compared to LCCO-

AV. HAN on batch reactor test with LCIO-AV gives a pressure rise of 1.9 bar. The ignition delay associated with HAN decomposition using the LCIO-AV catalyst is found to be in the range of 0.2–0.3 s, which is slightly lower than that observed for the LCCO-AV catalyst. The reduced ignition delay highlights the superior catalytic activity of LCIO-AV, enabling faster initiation of HAN decomposition.

Chapter 6 summarises the major outcomes of the thesis, emphasizing computational design of ionic liquids, HAN-IL fuel formulation, and catalyst development. It outlines the contributions of the work. Chapter 6 also proposes further avenues for research, including improved catalyst durability, development of advanced fuel formulations, and thruster-level demonstrations. Future studies aims to support the continued development of efficient, reliable, and practical green monopropellant systems.

CHAPTER 2

SYNTHESIS AND CHARACTERIZATION OF CONCENTRATED HYDROXYL AMMONIUM NITRATE (HAN)

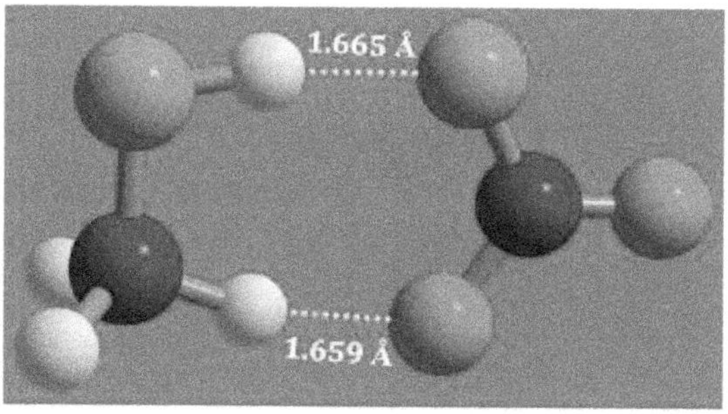

2.1 Introduction

Hydrazine (N_2H_4) has been widely used as a monopropellant owing to its high performance and the low temperatures required for catalytic decomposition in thrusters. Its classification as a potential human carcinogen, together with a relatively high vapour pressure at room temperature, creates significant challenges for safe storage and handling, ground processing and logistics, raising operational costs. Energetic ionic liquids are considered as promising alternative to replace the conventional propellant. The most studied ionic liquids for propulsion applications include hydrazinium nitroformate (HNF), hydroxylammonium nitrates (HAN), and ammonium dinitramide (ADN).

Hydroxylammonium nitrate (HAN) has garnered significant attention within the class of energetic ionic-liquid (EIL) monopropellants owing to its high energy density and comparatively benign post-combustion products (Werling et al., 2020; Gohardani et al., 2014). HAN exists as the hydroxylammonium cation (NH_3OH^+) paired with the nitrate anion (NO_3^-). This ionic pairing, rather than a neutral molecule is the defining structural feature that places it among ionic liquids. HAN-based ionic liquids known as 'green monopropellants' have low freezing point, high density, low environmental impact, and negligible vapour pressure. Additionally, HAN's higher density allows for a smaller storage tank on-board a spacecraft, thus reducing weight and cost. Hydrazine and HAN-based monopropellants have specific impulse (Isp) around 220 and 270 s, respectively, with the values of HAN tending to be slightly higher depending on the fuel components that make up the monopropellant composition.

HAN can be synthesised using various methods. The method for producing HAN by reacting hydroxyl ammonium sulfate (HAS, $(NH_2OH)_2 \cdot H_2SO_4$) and barium nitrate ($BaNO_3$) is simple but with lower yield and contains impurities (Liggett, 1978). Another method involves the electrolysis of nitric acid yields HAN with an excess of nitric acid which is undesirable. Another method involves the reaction of hydroxyl ammonium sulfate either with ammonia in ethanol to undertake ion exchange with nitric acid (Wagaman, 1990) or sodium nitrate

($NaNO_3$) in N-butyl alcohol (C_4H_9OH) (Liggett, 1993). In all of these methods, the final product may be of low purity or yield and requirement of catalyst.

The synthesis method selected influences the purity, yield and performance of the final HAN solution. The development of monopropellant formulations of HAN with energetic ionic liquids necessitates the standardization of a synthetic procedure to achieve concentrated HAN solution of consistent concentration and purity. In order to prepare HAN of high purity, the neutralization method from liquid hydroxylamine (HA, NH_2OH) and nitric acid that minimizes the formation of intermediates (Coates, 2000; Cawlfield, 1993) is selected. The synthesis, characterization and performance evaluation of 80 wt.% HAN solution, which is used for the preparation of HAN-energetic ionic liquid monopropellant formulations, is presented in this chapter.

2.2 Experimental

2.2.1 Materials

Hydroxyl amine (50% solution) was procured from Sigma Aldrich, USA. Analytical reagent grade 70% nitric acid was purchased from Merck India Ltd, Mumbai. The chemicals were used without any further purification. Deionized water was used for the preparation of HAN samples

2.2.2 Synthesis of Hydroxyl Ammonium Nitrate

HAN was synthesized in two steps. The first step involves the synthesis of dilute HAN and second step involves concentrating the synthesized HAN by vacuum distillation. In step one, 250 ml of hydroxyl amine solution was taken in a jacketed reaction vessel and cooled to ≤4°C. Nitric acid pre-cooled to 4°C was added to the hydroxyl amine solution in the jacketed vessel with stirring. A glass pH electrode was inserted into the solution to monitor the pH of the reaction mixture precisely. Since the reaction is exothermic the addition of nitric acid is controlled such that the temperature of the reaction medium does not exceed 4°C as measured using a thermometer. The end point of the neutralization was followed using the pH electrode. The reaction is considered complete when the pH of the reaction mass reaches 2.5. The concentration of the obtained aqueous HAN solution was estimated to be 40-45 wt.%.

The resulting HAN solution was further concentrated in a rotary evaporator under vacuum to yield concentrated HAN. The parameter used for concentrating dilute HAN are:

Vacuum pressure: 13 mbar

Bath temperature: 45-50°C

Chiller temperature: 5-6°C

RPM of flask: 150

The HAN synthesized was characterized for its purity and concentration by different analytical methods.

The synthesis of HAN up to 90 wt.% concentration was successfully carried out by the vacuum distillation method. However, for the thesis work, 80 wt.% HAN was employed in experimental studies due to safety considerations associated with laboratory-scale handling. For consistency and comparison, the theoretical studies were also carried out at the same concentration.

2.2.3 Characterisation of HAN

Assay Determination: The concentration of HAN was estimated by Potentiometric titration using an automatic titrator. A Metrohm 905 Titrando automatic Potentiometric titrator was used for the non-aqueous titrations. Titrations were carried out with approximately 0.1 molar tetrabutylammonium hydroxide, standardised with benzoic acid in denatured alcohol as titrant (Decker et al., 1988).

Chemical analysis: Hydroxyl amine is quantitatively oxidized to nitric acid by an excess of bromine monochloride. Aliquots of solution with 0.3 to 0.4 g of sample was prepared into a 250 ml volumetric flask and then make up with distilled water. 10 ml of the made-up solution was pipetted out into a clean flask and treated with 25 ml of KBr-$KBrO_3$ solution and 15 ml of concentrated HCl were added, stoppered and kept for completion of reaction. After 15 minutes, 15 ml 10

% KI solution was added to the flasks and the mixture was titrated against standard sodium thiosulphate (0.1 N) using starch as an indicator. Perform blank and duplicate determinations. Equivalent weight of hydroxyl amine is molecular weight/6.

The acid content was determined by titrating with standard sodium hydroxide solution using phenolphthalein indicator.

***Elemental analysis*:** CHN elemental analyser is used for the determination of carbon, hydrogen and nitrogen in the sample. The sample was combusted at 1000°C in an atmosphere of pure and dry oxygen, to convert carbon to carbon dioxide, hydrogen to H_2O and nitrogen to oxides of nitrogen. The combustion products were swept out of the combustion chamber by inert carrier gas such as helium and passed over heated high purity copper, to remove any oxygen not consumed in the initial combustion and to convert any oxides of nitrogen to nitrogen gas. The gases are then passed through the absorbent traps to leave only carbon dioxide, water, and nitrogen. Detection of the gases can be carried out by a GC separation followed by quantification of individual components as a function of their thermal conductivity using thermal conductivity detector. The hydrogen and nitrogen content of the HAN sample was determined using Elementar CHNS analyser.

Density: The density of HAN was measured using an Anton Paar DMA 35 portable digital density meter. The HAN solution sample was introduced into a U-shaped borosilicate glass that is being excited to vibrate at its characteristic frequency electronically. The characteristic frequency changes depend on the density of the sample. Determination of characteristic frequency enables to calculate the density of sample automatically.

***NMR analysis*:** Nuclear magnetic resonance spectroscopy (NMR) is used for the characterization of materials based on the magnetic property of atomic nuclei present in the sample. The NMR has a sample probe placed in a strong magnetic field and is irradiated with radiofrequency leading to nuclear level transitions. Depending on the atomic nuclei, the excitation frequency changes and the excited nuclei undergo free induction decay (FID) which is recorded. The FID generated in the time domain is converted to frequency domain by Fourier transformation resulting in the NMR spectrum output. NMR analysis of HAN was carried out in deuteriated water with a BRUKER AVANCE FT-NMR spectrometer. Inverse-gated proton decoupling method was applied for ^{15}N NMR. ^{15}N NMR chemical shift was referenced to 90% formamide at 112.7 ppm.

FTIR analysis: Fourier transform infrared spectroscopy (FTIR) is a fast and easy analytical technique for the identification of functional groups. It is based on the interaction of molecules with IR radiation. Certain frequencies characteristics of the vibrational energy levels are absorbed by the molecule and the rest is transmitted. Thus the spectrum will be characteristic of the chemical structure of the molecule. FTIR analysis was performed using Nicolet iS50 FTIR spectrometer equipped with variable temperature cell accessory. The sample was prepared by forming a thin film on a KBr pellet by smearing the HAN solution. The spectrum was recorded over the wavenumber range of 4000-400 cm^{-1} at a resolution of 4 cm^{-1}.

Raman analysis: Raman spectroscopy is based on the inelastic scattering of monochromatic light when the frequency of photons changes upon interaction with a sample. The photons of the laser light are absorbed by the sample and subsequently re-emitted. Frequency of the re-emitted photons is shifted up or down in comparison with the original monochromatic frequency, which is known as the Raman Effect. The Raman shift provides information about vibrational and rotational energies of molecular bonds. The Raman spectroscopy was performed using a WITec alpha 300R confocal Raman microscope using the 532 nm line of Nd: YAG laser as the excitation wavelength and charge coupled device (CCD) as the detector. An average of five spectra was collected with an integration time of 2 seconds/spectra to obtain a single spectrum plotted as intensity vs. wave number (Raman shift). The spectrometer was calibrated using the 520 cm^{-1} line of Silicon before analysis.

Inductively Coupled Plasma-Optical Emission Spectrometer (ICP-OES) analysis: Inductively Coupled Plasma Optical Emission Spectrometry (ICP-OES) utilises a high-temperature plasma to excite atoms and ions within a sample to emit light at element-specific wavelengths. This light is then separated into a high-resolution 2D spectrum by an Echelle polychromator and captured by a Charge Coupled Devices (CCD) detector, which allows for the simultaneous quantification of both major components and trace impurities. ICP-OES analysis was carried out using Thermo Scientific i-CAP 7000 series having a segmented array CCD detector to estimate the concentration of metallic ions in HAN solution.

A dilute aqueous solution (10-20 ppm) of the sample was used for the determination of trace metal ions in HAN solution. The concentration of Fe, Cu and Co in HAN samples was analysed and quantified from the calibration curve made from the standard solutions at 1ppm, 10 ppm and 20 ppm of the reference material under investigation.

***UV-Vis analysis*:** Ultraviolet (200-400 nm) and visible (400-800 nm) radiation are found towards the short wavelength of electromagnetic spectrum. ΔE is defined as the difference in energy between an occupied orbital (ground state) and an empty (excited state) orbital. When the energy of the incoming photon matches ΔE, the photon is absorbed, and an electron from an occupied level is excited from its ground state to an empty excited state which occurs in the UV-visible region of the electromagnetic spectrum. UV-Vis spectra tend to be broad in nature because vibrational and rotational levels of the molecular orbitals are superimposed upon the electronic levels. This broad nature makes their usefulness in identifying materials limited, but the technique is ideal for quantitative analysis of species in solution media. A UV–Vis spectrophotometric method for quantifying nitric acid in hydroxylammonium nitrate relies on the strong ultraviolet absorption of nitrate at 302 nm. The sample was appropriately diluted into the linear range, measured in a quartz cuvette against a reagent blank, and nitric acid concentration was determined from a calibration curve prepared from a nitrate (e.g., KNO_3) standards. The analysis was carried out using Agilent Cary 4000 UV-Vis spectrophotometer equipped with a deuterium arc lamp as source and R928 photomultiplier tube detector.

pH measurement: pH meter is used to accurately determine the acidic/basic nature of a solution. It consists of a pH sensitive glass electrode that responds to the activity of hydrogen ions in the sample and generates a potential difference proportional to the difference in concentration. Based on the Nernst equation, the potential is interpreted by the meter and converted into a pH value. Systronics 362 pH meter, having an accuracy of ±0.2 pH, calibrated with standard buffer solutions of pH 4, 7 and 9.2 was used.

Moisture content analysis: Water estimation was carried out using the Mettler Toledo Coulometric Karl Fischer titrator. The titration cell consists of the anodic and the cathode compartment and polarized double-platinum-pin electrode. The titrant used was Hydranal reagent and methanol was the titration solvent. 0.2-0.3 g of the sample was dissolved in methanol and titrated directly with the reagent. Karl Fischer titrations were carried out till the potential at the polarized double-platinum-pin electrode falls below 100mV which is considered as the endpoint.

Thermo gravimetric analysis (TGA): The thermal decomposition characteristics of HAN samples were studied using differential thermal analysis-thermogravimetric analysis (TA instrument-Q50). Thermogravimetric analysis (TGA) measures the mass change associated with materials as a function of temperature and time. To understand the decomposition behaviour, a decomposition kinetics analysis of the HAN was carried out using the model-fitting Kissinger method. Different heating rates were chosen for the Kissinger method (Kissinger, 1957) and the activation energy E_a is attained from Eq. (2.1).

$$\ln\left(\frac{\beta}{T^2}\right) = \ln\left(\frac{AR}{T}\right) - \frac{Ea}{RT} \qquad (2.1)$$

where β is the heating rate, T is the temperature in Kelvin scale at a fixed conversion value of α. A is the pre-exponential factor, E_a is the activation energy at a given conversion and R is the gas constant. For the Kissinger method, $\ln\left(\frac{\beta}{T^2}\right)$ was plotted against $1/T$ giving rise to a straight line whose slope yields E_a.

TG/MS analysis in Perkin Elmer Pyris 1 TGA coupled with Clarus SQ8T quadruple mass spectrometer was used for the identification of evolved species from 10 to 500 amu mass range in He atmosphere at a heating rate of 5 °C/min.

***DSC analysis*:** In Differential Scanning Calorimetry (DSC), the differential heat flow associated with material phase transitions is measured as a function of time and temperature. The sample and reference materials are subjected to a precisely programmed furnace. The temperature of the furnace is raised linearly and the resultant differential heat flow to the sample and reference is monitored by thermocouples and recorded against temperature or time. The properties such as kinetics, melting point and purity, phase transition and glass transition can be evaluated from DSC. Glass transition temperature was determined using a differential scanning calorimeter (DSC Q20, TA instrument, USA) in the N_2 atmosphere. In order to measure the glass transition temperature, the HAN sample was first cooled from +50°C to -150 °C and then heated from -150 °C to +50°C, at the rate of 5 °C/min.

Batch reactor: The batch reactor comprises a small stainless steel (SS-304) chamber in which the pressure and temperature rise are monitored following monopropellant injection. The reactor chamber was maintained isothermally at 175 °C using a PID-controlled heater mounted beneath the bottom plate. During pulse-mode operation, approximately 200 μL of HAN was introduced into the preheated chamber, and the resulting pressure rise was recorded using pressure transducers to evaluate the decomposition behaviour of the monopropellant.

Computational/Theoretical Methodology: The DFT calculations were performed with B3LYP using Gaussian 09 software. The basis set used was 6-311+G(d,p). In the study, atomization data obtained from the literature along with the G3MP2 level thermodynamic data was used to calculate the heat of formation. The G3MP2 is one of the popular high accuracy methods for calculating enthalpy quantities which is based on complex energy computations involving several pre-defined calculations on the specified molecular system (Curtiss et al. 1999).

The theoretical prediction of the propellant performance was carried out by equilibrium combustion analysis using the NASA Chemical Equilibrium with Applications (CEA) computer code (Frem, 2018), for a chamber pressure (Pc) of 20 bar and an area ratio (Ae/At) of 50 for oxidizer to fuel ratios ranging from 1.0 to 7.0. Additional inputs to the model were heat of formation (determined theoretically) and the temperature of the reactants (298.15 K). The theoretical performance parameters include specific impulse($Isp_{(vac)}$) and combustion temperature (Tc).

2.3 Results and Discussion

Hydroxylammonium nitrate (HAN) is synthesized through the controlled neutralization of hydroxylamine (HA) with nitric acid, resulting in a high purity product devoid of secondary by-products or impurities, as represented in the **Fig. 2.1**

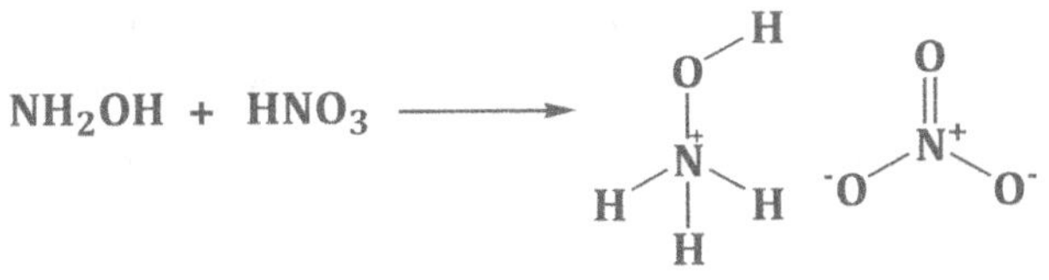

Figure 2.1 Scheme for the synthesis of HAN by neutralization reaction.

Fig. 2.2 is a schematic of the experimental setup for synthesis of HAN. The reaction is highly exothermic and is therefore carried out at a temperature below 4 °C under vigorous stirring, with nitric acid added slowly to the hydroxylamine solution. Maintaining a low temperature is essential because, if the reaction mixture exceeds this temperature limit, hydroxylamine undergoes thermal decomposition to form ammonia. The liberated ammonia can then react with nitric acid, leading to the undesired formation of ammonium nitrate as an impurity in the final product. Furthermore, to minimise the risk of self-decomposition of the HAN formed, 40% nitric acid was employed. This controlled concentration ensures effective protonation of hydroxylamine, since it is a weak base, while nitric acid acts as a strong acid to stabilize the salt.

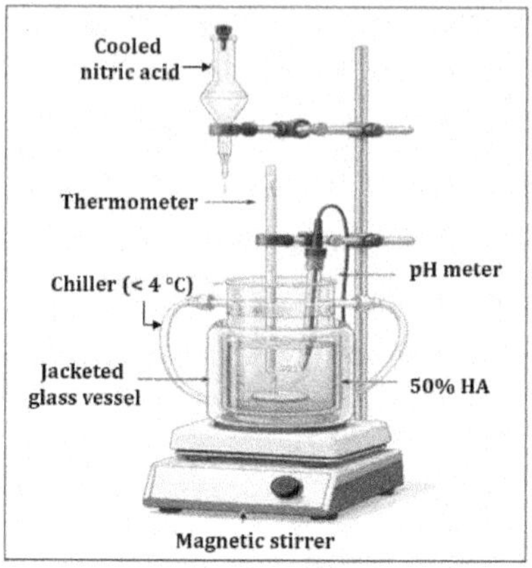

Figure 2.2 Experimental set up for synthesis of HAN.

pH of the 110 ml of 50 wt.% hydroxyl amine solution as a function of volume of 40% nitric acid added is presented in **Fig. 2.3**. The HA solution shows an initial pH of 10.5. The HA solution requires 215 ml of 40% nitric acid to reach a pH of 2.5, which is considered as the endpoint of the reaction. The reaction product at the equivalence point (pH 2.5) is HAN solution of approximately 45 wt.% concentration. This low concentration may pose difficulties in the effective use of aqueous hydroxylammonium nitrate in propellant applications. Hence it is further concentrated using a rotary vacuum evaporator. Removing water from aqueous HAN becomes difficult as the HAN concentration increases. HAN concentration more than 80 wt.% is achieved by increasing vacuum (10 mbar) and temperature (55°C) conditions.

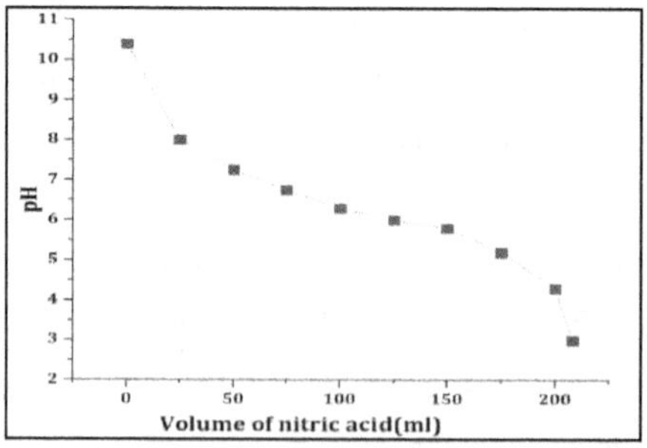

Figure 2.3 Variation of pH with the volume of 40 % added nitric acid in HAN synthesis.

2.3.1 Assay determination

The assay determination of HAN requires quantifying the concentrations of hydroxylammonium, water, and nitrate ions present in the sample. Titrimetric methods provide a reliable and accurate analytical approach for establishing the assay values of these components. The hydroxylammonium ion content, determined using a standard iodometric titration procedure is equivalent to a HAN concentration of 80 wt.%. **Fig. 2.4** represents the potentiometric titration curve used for determining the HAN concentration. In this method, titration performed in denatured alcohol as the solvent medium provided a clear and well-defined inflection point, enabling precise endpoint identification. The potentiometric titration yielded a HAN concentration of 80.5wt.%, demonstrating good agreement with the iodometric assay value and confirming the reliability of the potentiometric technique for accurate HAN quantification.

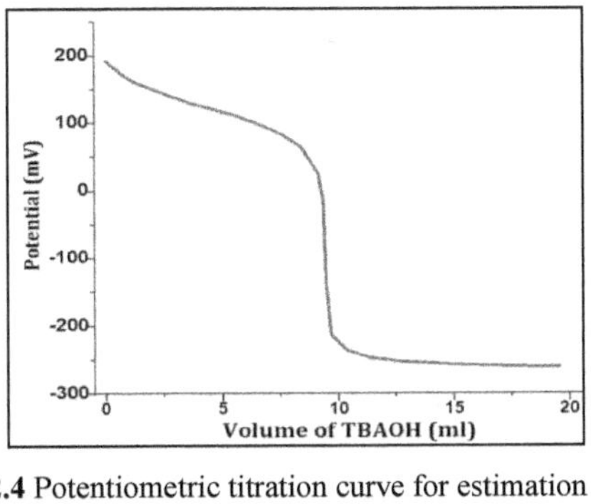

Figure 2.4 Potentiometric titration curve for estimation of HAN.

The density depends on the concentration of HAN solution and temperature. A higher density propellant is desirable because it allows a greater mass of propellant to be stored within the same tank volume, thereby improving volumetric efficiency. This in turn, contributes to a reduction in the overall weight and size of spacecraft storage tanks, offering significant advantages in propulsion system design. The density of 80wt.% HAN solution determined at different temperatures (density of water at 20°C=0.999 g/cc) is shown in **Fig. 2.5(a)**. The density of 80 wt.% HAN solution varies from 1.5778 g/cc to 1.5741 g/cc when the temperature increases from 20 to 25 °C. The density of aqueous HAN solutions of different concentrations determined experimentally at a temperature of 25 °C, is presented in **Fig. 2.5(b)**. The solutions of lower concentrations for density measurement are obtained by diluting the prepared 80 wt.% HAN solutions using deionized water. The density of HAN solution at 25 °C varies from 1.2611 to 1.5743 g/cc when the concentration increases from 43.9 to 80 wt.%.

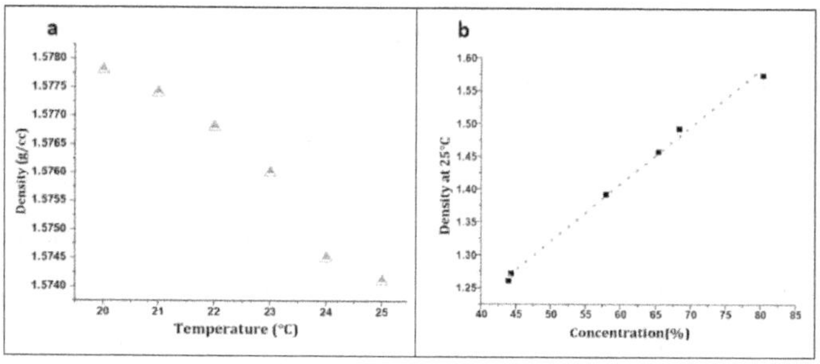

Figure 2.5 Plot of (a) Density vs Temperature of 80 wt.% HAN solution and (b) Density vs Concentration of HAN solution measured at 25 °C.

The presence of water in HAN plays a crucial role in governing its stability and energetic performance. Water acts as a stabilizing component in the HAN matrix by moderating ion–ion interactions, reducing viscosity, and influencing decomposition phenomenon. The water content in HAN can be determined indirectly by knowing the concentration of HAN from titrimetric analysis. However, accurate determination of water content is done using coulometric Karl Fischer titration. The Karl Fischer titration yielded a water content of 17-18%.

The CHN analysis is performed to confirm the purity of the synthesized HAN. The elemental composition obtained by the CHN analysis of 80% HAN agrees well with the theoretical values. The elemental composition obtained is: H-4.6%, N-24.5% vis-à-vis the theoretical values of H-4.7%, N-24.7%.

HAN solution is highly sensitive to changes in pH and the presence of metal-ion impurities (iron, copper, etc.). Iron initiates a reaction sequence for HAN that involves the reduction of Fe(III), followed by the oxidation of Fe(II) by nitrate, generating nitrous acid, which further reacts with HAN, causing an autocatalytic reaction. As the concentration of iron increases, the temperature required to initiate the decomposition reaction decreases. Harlow et al. (1998) reported that the autocatalytic temperature decreases from 74°C to 42°C as the iron concentration increases from zero to 1874 ppm. Moreover, the metal ions can undergo redox reactions, producing gases that result in a gradual pressure build-up in sealed system. Acidic environments further accelerate autocatalytic decomposition producing ammonium nitrate, nitric acid and dihydroxylamine as by-products. Hence it is highly essential to determine the metal ion content in HAN solution. The purity of the HAN solution is assessed by quantifying trace metal ions using ICP-OES analysis. The ICP results indicated that the concentrations of Fe, Co and Cu are below 1 ppm, confirming that the HAN sample has a high level of purity.

2.3.2 Spectroscopic analysis

Spectroscopic methods such as Raman, FTIR and ^{15}N NMR confirm the successful formation of HAN. The ^{15}N NMR confirm the successful formation of HAN with the two type of N atoms as shown in **Fig. 2.6.** The signals observed at 375.6 ppm and 83.6 ppm are arising from the N atom of nitrate anion and hydroxyl amine cation, respectively.

The nitrogen resonance associated with free hydroxylamine appears at 104.4 ppm, however this signal shifts to 83.6 ppm in hydroxyl ammonium ion. The upfield shift of quaternary nitrogen from free hydroxyl amine might be due to changes in electronic environment owing to protonation and hydrogen bonding. Upon protonation the lone pair is converted to N-H sigma bond leading to reduction in lone pair related paramagnetic contributions that significantly reduces deshielding. Further strong symmetric hydrogen bond present in hydroxyl ammonium increase shielding at nitrogen. Both of these effects lead to an upfield resonance of the Hydroxyl ammonium N. The resonance signal of nitric acid at 356.5 ppm is shifted to 375.6 when it forms an ion pair with hydroxylamine.

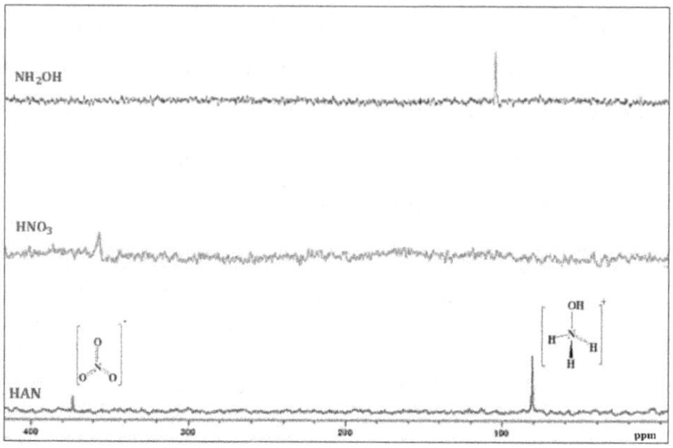

Figure 2.6 ^{15}N NMR spectrum of HA, Nitric acid and HAN.

The FTIR spectrum of 80% HAN has vibration bands at 3174, 2924, 1767, 1517, 1175, and 1004 cm^{-1} assigned to the NH_3OH^+ group. The band at 2715 cm^{-1} is the combination band arising from the rocking and symmetric deformation of HA^+. The N–O bands of NO_3^- have a doubly degenerate peak at 1338 cm^{-1} and 1385 cm^{-1} and a deformation band at 824 cm^{-1}. The stretching vibrations of the O–H bond are identified at 3450 and 1622 cm^{-1}.

The FTIR of HAN has clear concentration-dependent spectral changes showing ion-pairing/ hydrogen bonding and solvent interaction. The FTIR of 80% HAN is compared with 85% HAN in **Fig. 2.7.** The band assigned to hydroxyl ammonium ion (1767 cm^{-1}) and N-O deformation of nitrate ion (1045 cm^{-1}) is relatively more intense when the concentration of HAN is 85% due to increased ion pair interactions. The low frequency nitrate deformation peaks at 824 cm^{-1} and 722 cm^{-1}corresponding to the bonded nitrate ion is slightly more intense when the concentration is increased. Moreover the vibrations of the O–H bond are relatively weakened at 85% HAN. Hence, increasing the HAN concentration strengthens the hydrogen bonding interactions and ion pairing between hydroxyl ammonium cation and nitrate anion, which can influence solution properties such as ionic mobility, stability and decomposition reaction.

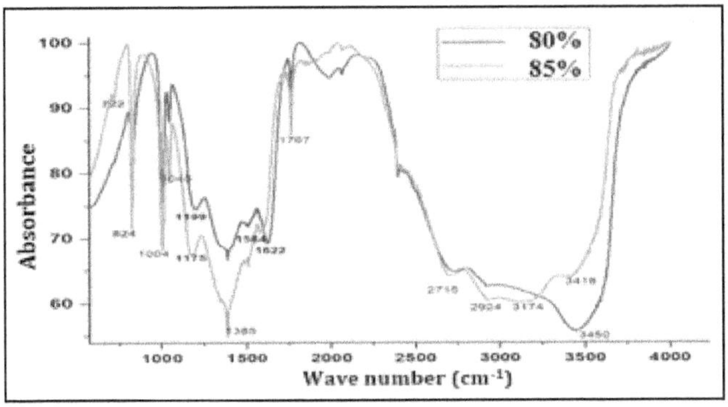

Figure 2.7 FTIR spectrum of 85wt.% and 80wt.% HAN solutions.

Fig. 2.8 shows the Raman spectrum of 80 wt.% aqueous HAN solution. The peak corresponding to the N-O symmetric stretching vibrations of hydroxylammonium ion is located at 1001 cm^{-1}. The peak associated with the NH_3 asymmetric deformation mode of HA^+ is assigned at 1589 cm^{-1}. The peak in the region of 2700-3000 cm^{-1} corresponds to the signature vibration modes of the hydroxyl ammonium cation. The spectrum is characterized by the symmetric N-O stretching vibrations of NO_3^- at 1040 cm^{-1} and anti-symmetric stretching to 1410 cm^{-1}. A band at 712 cm^{-1} is associated with the in-plane bending mode vibration of NO_3^-. The water molecule in aqueous HAN has three Raman active modes. The O-H stretching, O-H bending and O–H anti-symmetric stretching at 3143, 1634 and 3486 cm^{-1}, respectively. The peak located at 1185 cm^{-1} is due to the presence of bonded hydroxyl group (-OH) in aqueous HAN.

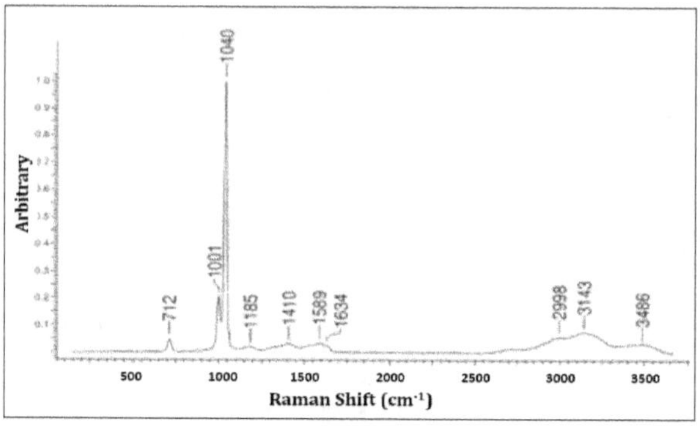

Figure 2.8 Raman spectrum of 80 wt.% aqueous HAN solution.

The nitrate ion concentration in HAN is quantified using UV–Visible spectroscopy. Nitrate ions are strongly UV-active and exhibit characteristic absorption bands at approximately 206 nm and 302 nm (Meyerstein and Treinin, 1961). Although the 206 nm transition shows higher molar absorptivity, it also exhibits a saturation effect at elevated nitrate concentrations, reducing its analytical reliability for concentrated HAN solutions. Therefore, the absorbance at 302 nm is selected for quantitative analysis. **Fig. 2.9** gives the absorption spectrum of HAN, hydroxyl amine and nitric acid. The nitrate concentration in HAN samples is determined by interpolating the measured absorbance values on a calibration plot obtained using a standard KNO_3 solution.

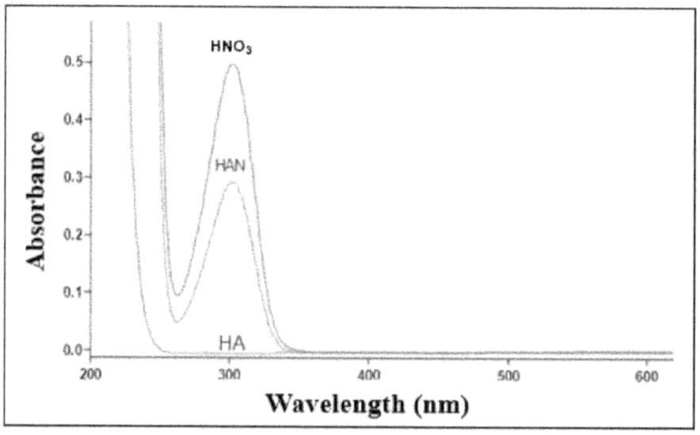

Figure 2.9 UV-Vis absorption spectrum of HAN, HA and HNO_3.

To validate the spectroscopic results, the UV data are compared with values obtained from a standard acid–base titration using NaOH. The comparison showed good agreement between the two methods, confirming the reliability of the UV–Vis method for nitrate estimation. However, for determining the HAN concentration, particularly in samples containing excess nitric acid, UV-Vis measurements may not differentiate between nitrate from HAN and free nitric acid contamination. The nitrate content in HAN solutions of various concentrations estimated by the UV-Vis spectroscopic method is compared with the theoretical value in **Table 2.1.**

Table 2.1 Nitrate content in HAN with different concentrations.

HAN concentration (wt.%)	Nitrate content (%)		
	Theoretical	Chemical method	UV-Vis method
58.1	37.52	37.84	38.56
80	51.7	51.78	51.73

2.3.3 Thermal characterization

The thermal decomposition of 80 wt.% HAN solution has been studied using thermogravimetry (TG), differential scanning calorimetry (DSC), a batch reactor and evolved gas analysis using mass spectrometry coupled with TG. The low temperature behaviour of HAN solution is analysed by low temperature DSC. The DSC analysis exhibits an endotherm at -100 °C corresponding to the melting point of frozen 80 wt.% HAN solution indicating its wider liquid range. The DSC graph of 80wt.% HAN solution is shown in **Fig. 2.10.**

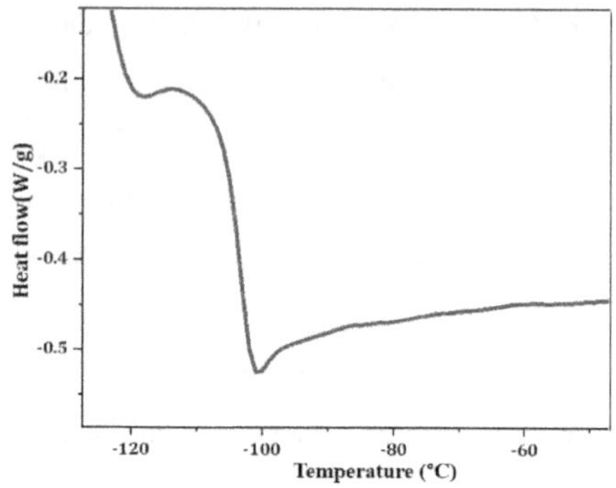

Figure 2.10 Low temperature DSC plot for 80wt.% HAN solution.

The TGA-DTG graph of aqueous HAN solution is shown in **Fig. 2.11(a).** The slow weight loss observed at temperatures up to 100 °C is due to the evaporation of water, which results in an increase in concentration of HAN in the solution. A rapid weight loss observed at temperature in the range of 165 to 220 °C is due to the thermal decomposition of HAN. The DTG graphs indicate the onset of thermal decomposition at 158±2°C and completion of decomposition at 220 °C. The peak decomposition temperature observed is 195 °C.

The DSC graphs show a broad endotherm below 100 °C corresponding to the evaporation of water. A second major sharp exotherm observed at 155 °C corresponding to the thermal decomposition of HAN (Amrousse et al., 2012). The DSC graph for the thermal decomposition of 80 wt.% HAN solution is presented in **Fig.2. 11(b).**

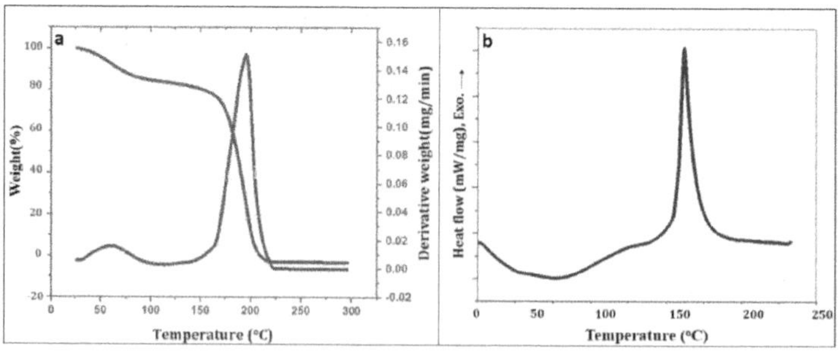

Figure. 2.11 (a) TG-DTG and (b) DSC plot of 80wt.% HAN solution.

The thermal decomposition products are detected using TG-MS. **Fig. 2.12** shows the most relevant mass fragments (m/z) produced during the thermal decomposition of HAN. The analysis of evolved gases using a mass spectrometer coupled with the TGA reveals peaks corresponding to the molecular mass (m/z) of 30, 44, and 28, indicating the formation of NO, N_2O and N_2 molecules during the thermal decomposition of HAN. Thermal decomposition of HAN commences only after the complete evaporation of water from the solution.(Chai et al., 2022). Due to the gradual release of water, it is usually not detected in the mass spectrometer.

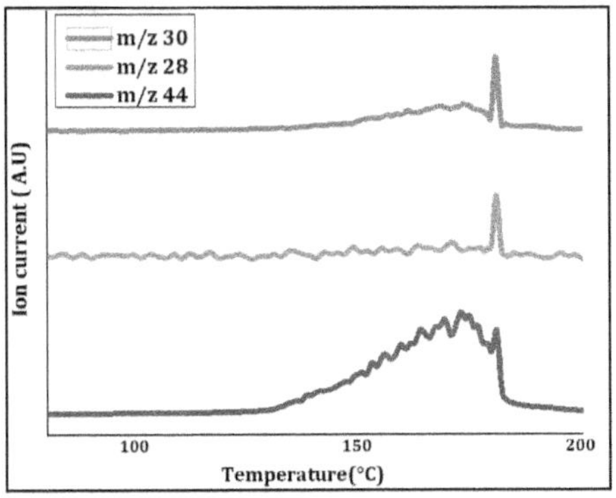

Figure. 2.12 Evolved gases during the thermal decomposition of HAN.

The thermal decomposition of hydroxylammonium nitrate (HAN) proceeds by a proton transfer to form neutral hydroxylamine and nitric acid, which subsequently produce HNO_2 and HNO. Hydroxylamine reacts with nitrous acid to yield nitrous oxide (N_2O; m/z 44) and water. Reaction of hydroxylamine and HNO gives molecular nitrogen (N_2; m/z 28). Nitrous acid also decomposes into NO (m/z 30). Secondary reactions such as the conversion of HNO_3 and HNO into HNO_2, or the disproportionation of nitrous acid into NO and HNO_3, support these gas-phase products, illustrating the multistep thermal decomposition behaviour of HAN as reported earlier (Lee and Litzinger, 2003). The different steps involved in the thermal decomposition of HAN to generate gaseous products such as NO, N_2O and N_2 are given in Eqs. (2.2) - (2.8).

$$NH_3OH^+ NO_3^- \leftrightarrow NH_2OH + HNO_3 \quad (2.2)$$

$$NH_2OH + HNO_3 \rightarrow HNO + HNO_2 + H_2O \quad (2.3)$$

$$NH_2OH + HNO_2 \rightarrow N_2O + 2H_2O \quad (2.4)$$

$$NH_2OH + HNO \rightarrow N_2 + 2H_2O \quad (2.5)$$

$$HNO + HNO_3 \rightarrow 2HNO_2 \quad (2.6)$$

$$2HNO \rightarrow N_2O + H_2O \quad (2.7)$$

$$3HNO_2 \rightarrow 2NO + HNO_3 + H_2O \quad (2.8)$$

The kinetics of HAN decomposition is studied from the Kissinger plot using the peak temperature (Tpeak) in DSC analysis. The decomposition is carried out at heating rates (β) of 1, 2, 2.5 and 3 °C/min. The activation energy of decomposition determined from the slope of ($\ln(\beta/T^2)$) vs 1/T plot is 63.2 kJ/mol. The activation energy obtained agrees with the literature data of 64.28 kJ/mol(Esparza et al., 2018). The Kissinger Plot of thermal decomposition of 80 wt.% HAN is shown in **Fig. 2.13.**

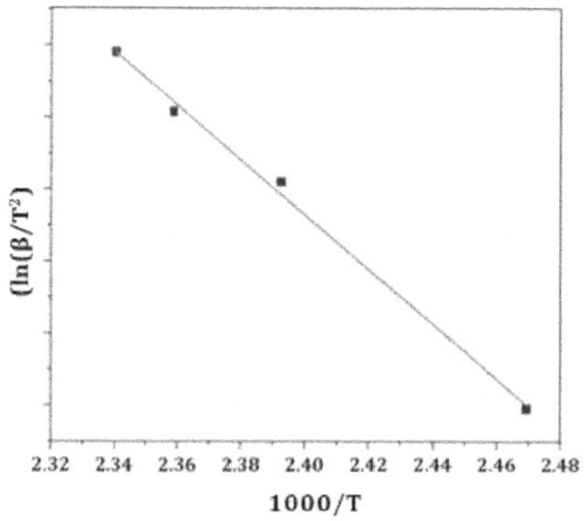

Figure 2.13 Kissinger Plot ($\ln(\beta/T^2)$ vs 1000/T) for thermal decomposition of 80 wt.% HAN.

The batch reactor, designed and built in-house is used for HAN decomposition studies. A schematic of the batch reactor test setup and its photograph are shown in **Fig. 2.14(a)** and **Fig. 2.14(b)**, respectively. In pulse mode operation, approximately 200 μL of HAN is injected into a chamber preheated to fixed temperature. The pressure and the temperature rise inside the chamber are recorded over time. The pressure-time profile of HAN decomposition from the batch reactor test is given in **Fig. 2.15**. Batch reactor testing provides information on the rate of decomposition by following the pressure rise when HAN is injected onto a preheated chamber. The absolute pressure rise assesses the decomposition extent, as the formation of gaseous products directly contributes to the observed pressure increase. HAN decomposes to give a pressure rise to the maximum value of 0.9 bar, with an ignition delay, defined as the time needed for the chamber pressure to rise to 10% of the steady pressure value (Amrousse et al., 2013), of 14.5 s.

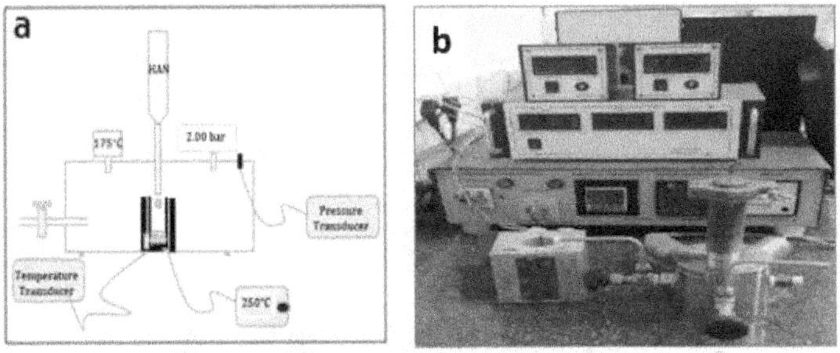

Figure 2.14 (a) Schematic of batch reactor test setup (b) Photograph of batch reactor.

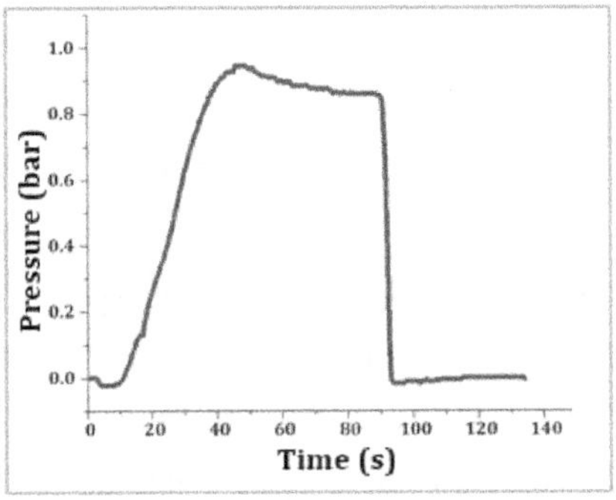

Figure 2.15 Pressure profile on 80wt.% HAN decomposition from batch reactor test.

2.4 Computational and Theoretical Studies

The theoretically optimised geometry of the isolated HAN molecule is obtained using DFT with B3LYP/6–311+G(d,p) basis set within the Polarizable Continuum Model (PCM) to determine the influence of medium-induced polarization on the ion-pair configuration. The optimised structure given in **Fig. 2.16** shows that the hydroxylammonium cation (NH_3OH^+) and nitrate anion (NO_3^-) form a strong ion pair, stabilised by two highly directional hydrogen bonds. The two hydrogen bonds at distances of 1.665 Å and 1.659 Å correspond to NO-H···O and N-H···O, respectively. These hydrogen bonding interactions control the stability and reactivity of HAN in the bond cleavage associated with the decomposition reactions.

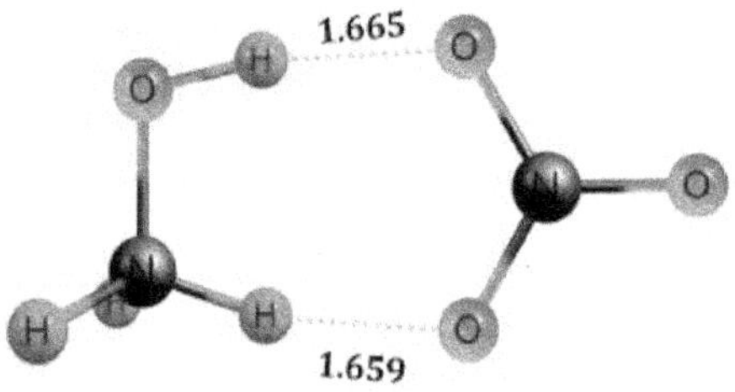

Figure 2.16 Optimized structure of HAN molecule.

Theoretical performance evaluation of HAN is carried out using NASA CEA computer code with heat of formation of HAN (-198.2 kJ/mol) calculated by the theoretical ab initio method (Thomas et al., 2015). Specific impulse (Isp $_{(vac)}$) is an important performance parameter which depends on the concentration of HAN. Theoretical Isp of 95 wt.% HAN and 80 wt.% HAN solution are 235 and 205s, respectively. That is, 95wt.% HAN shows Isp 30 units higher than that of 80 wt.% HAN along with augmented chamber temperature >1650 °C. Further, 80 wt.% HAN solution gives an Isp of 205 s which is lower than the corresponding value of hydrazine. It appears that Isp of HAN is limited by its oxidizer rich composition. Therefore, addition of a fuel in HAN solution is necessary to further enhance its performance. **Table 2.2** gives the performance comparison for hydrazine and HAN solutions.

Table 2.2 Theoretical Performance Comparison of hydrazine and HAN solutions.

Propellant	$Isp_{(vac)}$ (s)	Temperature (°C)
Hydrazine	230	900
95% HAN	235	1664
80% HAN	205	1170

2.5 Conclusions

The green monopropellant hydroxylammonium nitrate (HAN), a potential replacement for the state-of-the-art monopropellant hydrazine, is synthesized using the neutralisation method and concentrated by vacuum distillation. The final purity and concentration of HAN are determined by different chemical and instrumental methods. The highest concentration of ~ 80 wt.% HAN achieved has a density of 1.57778 g/cc at 20 °C. ^{15}N NMR spectrum exhibits two distinct peaks corresponding to the two different types of N atoms in HAN. The HAN is thermally stable up to 158°C with an exothermic decomposition at Tpeak of 195°C. The batch reactor test in pulse mode operations gives a pressure rise to the maximum value of 0.9 bar with an ignition delay of 14.5 s. Theoretical studies using NASA CEA indicate that 80 wt.% HAN yields an $Isp_{(vac)}$ of about 205 s. This Isp is limited by oxidizer rich composition and adding a fuel or fuel-rich additive is necessary to improve overall performance.

CHAPTER 3

HYDROXYL AMMONIUM NITRATE – 1-BUTYL-3-METHYL-IMIDAZOLIUM BASED MONOPROPELLANT: ROLE OF ANIONS IN ENHANCING IONIC LIQUID ENERGETICS

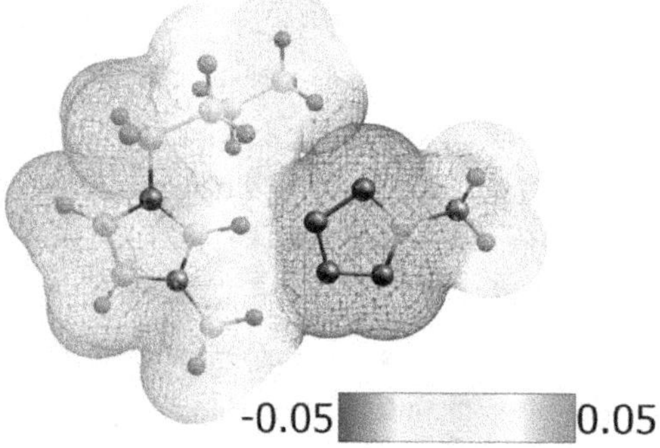

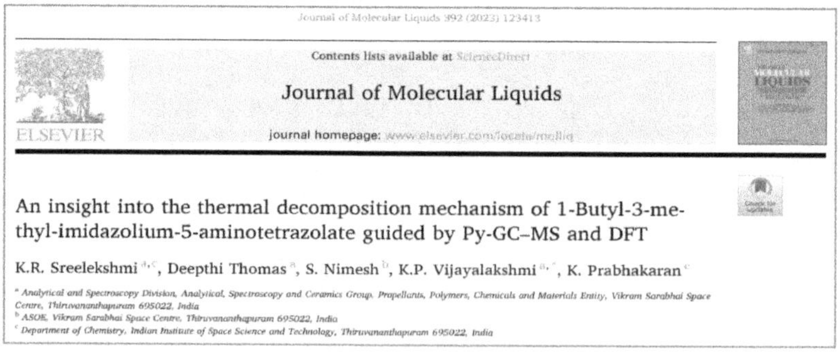

Journal of Molecular Liquids 392 (2023) 123413

Contents lists available at ScienceDirect

Journal of Molecular Liquids

journal homepage: www.elsevier.com/locate/molliq

An insight into the thermal decomposition mechanism of 1-Butyl-3-methyl-imidazolium-5-aminotetrazolate guided by Py-GC–MS and DFT

K.R. Sreelekshmi [a,c], Deepthi Thomas [a], S. Nimesh [b], K.P. Vijayalakshmi [a,*], K. Prabhakaran [c]

[a] Analytical and Spectroscopy Division, Analytical, Spectroscopy and Ceramics Group, Propellants, Polymers, Chemicals and Materials Entity, Vikram Sarabhai Space Centre, Thiruvananthapuram 695022, India
[b] ASOE, Vikram Sarabhai Space Centre, Thiruvananthapuram 695022, India
[c] Department of Chemistry, Indian Institute of Space Science and Technology, Thiruvananthapuram 695022, India

3.1 Introduction

Hydroxylammonium nitrate (HAN) acts both as an oxygen-rich oxidiser and a fuel component, yielding a net positive oxygen balance. To approach stoichiometry and enhance performance, HAN-based monopropellants are typically blended with fuels. Appropriately formulated HAN-fuel monopropellants can achieve improved specific impulse values approaching those of bipropellant systems while generating high adiabatic combustion temperatures. The fuel can be ionic or molecular. Selection of a fuel depends on its heat of formation, density, compatibility, stability, solubility, low toxicity and volatility (Wucherer et al., 2000). Most of the studies on HAN formulation is based on blends of HAN with hydrocarbon fuels for chemical propulsion applications (Chang et al., 2001; Kasbi et al., 2025; Amrousse et al., 2017). Currently, a new class of HAN propellant with imidazolium-based ionic liquid has been proposed as dual mode propulsion system (Berg and Rovey, 2012). Imidazolium-based ionic liquid fuel-HAN binary monopropellant mixture exhibits 1-4% improved specific impulse, exceeding the current state-of-the-art propellants (Berg and Rovey, 2013; Donius and Rovey, 2011).

Ionic liquids (ILs) are ion pairs usually composed of an organic cation and an organic/inorganic anion possessing unique properties. The non-solvent applications of ILs are recognized as advanced functional materials with the advantage over molecular materials with a wide choice of cations or anions, different alkyl substituents, and functional groups(Greer et al., 2020; Zhang et al., 2016). ILs' unique advantage relies on selecting a proper cation and anion to obtain a compound with desired properties, which directly translates into potential applications in various technological processes. ILs are unique in the sense that, in addition to ionic and covalent interactions, there exist weak H-bonding and π-stacking interactions. Energetic ionic liquids (EILs) are a special class of ILs that can thermodynamically provide high energy on combustion or decomposition. EILs have attracted much attention as environmentally benign explosives or propellant fuels (Sebastiao et al., 2014) due to their good thermal stability, excellent energetic performances, low volatility, wide liquid range, and reduced sensitivity. The characteristics of low viscosity, and high density associated with high positive heat of formation further add to the credit of EIL for propellant application. Some of the EILs are also explored for replacements of hydrazine in propellant formulations (Jin et al., 2022; Zhang and Shreeve, 2013). The presence of N-heteroatomic ions impart advantages such as high heat of formation arising from energy-rich N-N and C-N bonds (Jones et al., 2006) and produces safe dinitrogen as the main decomposition product. The designing of ILs based on nitrogen-rich heterocycles is expected to combine the virtues of traditional energetic materials with the core characteristics of ILs.

The imidazolium ring is chosen as the cation owing to its exceptional thermal stability, facile synthesis procedures (Wang et al. 2010), chemical stability (Weingarth et al. 2012) and fluid properties. The distinctive properties of imidazolium cations arise from the aromatic electronic structure. The N1-C2-N2 fragment in imidazolium cation exhibits a three-centre, four-electron bonding interaction and a double bond between C3 and C4. The C2 is positively charged owing to the electron scarcity in the C=N bond, imparting slight acidity to the C2-H hydrogen atom (**Fig. 3.1**). Hence the C2-H…anion interactions play an important role in the overall chemical and physical properties of the compound. The fluid properties of imidazolium based ILs which is essential for propellant application can be explained with this C2-H…anion interactions (Ludwig and Paschek, 2009). The major classes of imidazolium based ILs include 1-alkyl-3-methylimidazolium ($[RMIm]^+$), 1-alkyl-2,3-dimethylimidazolium ($[RMMIm]^+$) and symmetrical 1,3-dialkylimidazolium ILs ($[RRIm]^+$). Symmetrical ILs ($[RRIm]^+$ $[X]^-$) tend to pack more effectively in the solid state, and form salts with higher melting points. However, in monopropellant applications a lower melting point is preferred for ionic liquids. Hence, the cation with a reduced symmetry (1-butyl-3-methylimidazolium) is favored.

Task-specific ionic liquids can be designed by changing the N-alkyl substituent on the ring or pairing with different anions. The anion structure often has a prominent influence on the physico-chemical characteristics such as liquid range, melting point, electron density, and thermal behaviour, which arises due to the cation-anion interaction energies in the IL (Ofem et al., 2023, Sreelekshmi et al., 2023). The choice of anions often influences the stability and performance of EILs. While oxygenated anions are commonly used (Jones et al., 2006), they can sometimes lead to stability issues despite their low melting points. However, nitrogen-rich heterocyclic anions, such as azole-type anions, are highly promising for EIL design due to their charge-delocalized structures, high nitrogen content, and positive heat of formation, which can form stable ionic liquids with imidazolium cations. The work presented in this chapter mainly focuses on 1-butyl-3-methylimidazolium based ILs with five different anions as fuel for HAN-based monopropellant. The study enables exploration of EILs with butyl-substituted imidazolium cation and energetic anions as new materials for propellant fuel applications.

3.2 Experimental

3.2.1 Materials

5-amino tetrazole (97%), potassium carbonate (99%), and 1-butyl-3-methyl-imidazolium chloride (≥98%) were procured from Sigma-Aldrich and used as received. The analytical reagent grade acetone used as solvent was procured from Merck India Ltd, Mumbai. Distilled water was used for the preparation of solutions.

3.2.2 Computational methodology

The work theoretically investigates the energetic characteristics of novel ILs based on 1-butyl-3-methyl imidazolium cation paired with five different anions: nitrate (N), dinitramide (DN), 5-amino-1,2,4-triazolate (Tz), 3,5-diamino-1,2,4-triazolate (DTz) and 5-amino tetrazolate (AT) using Density functional theory. The cation/anion combinations of the ionic liquids studied in this work are given in **Fig. 3.1.**

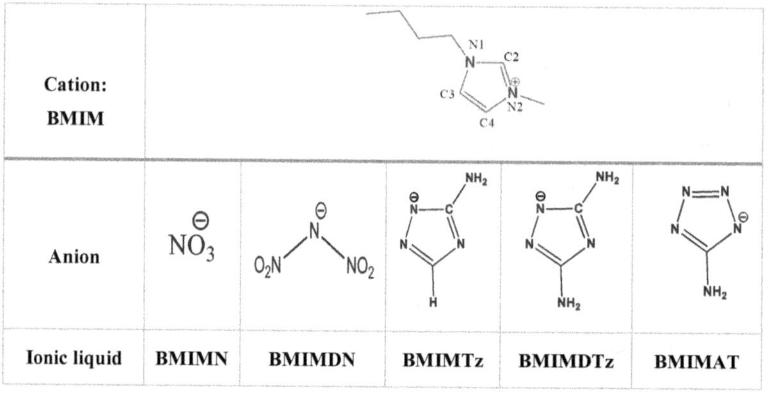

Figure 3.1 Chemical structures of BMIM cation and the selected counter anions.

Density functional theory (DFT) calculations were performed on each ion pair to obtain the optimized molecular geometries using a 6–311+G(d,p) basis set as implemented in Gaussian 09 software (Jiang et al., 2019; M.J. Frisch et al., 2009). The decomposition mechanism of IL was studied theoretically by B3LYP at 6-311+G(d,p) level of DFT. The structure of the transition state was confirmed by the presence of a single imaginary frequency and intrinsic reaction coordinate (IRC) calculations were carried out for the minimum energy pathway (Maeda et al., 2015). The heat of formation for individual ionic liquids was determined using the G3MP2 ab initio method.

3.2.3 Theoretical performance evaluation

The theoretical prediction of the propellant performance was carried out by equilibrium combustion analysis using NASA Chemical Equilibrium with Applications (CEA) computer code (Frem, 2018). Mixtures of EILs (fuel) with oxidizer (HAN) have been investigated in different oxidizer-to-fuel ratio under equilibrium conditions to determine the optimal high-performance monopropellant composition. HAN was chosen as the oxidizer due to very good solubility characteristics and positive oxygen balance (Berg and Rovey, 2013). The theoretical performance parameters include specific impulse ($Isp_{(vac)}$) and combustion temperature (Tc) at various oxidizer to fuel weight ratios (O/F=1 to

7), assuming a chamber pressure of 20 MPa and a nozzle expansion to atmospheric pressure and the area ratio, A_e/A_t=50. Additional inputs to the model were the heat of formation (determined theoretically) and reactant temperature of 25°C.

3.2.4 Synthesis of 1-butyl-3-methyl imidazolium-5-amino tetrazolate (BMIMAT)

1-butyl-3-methyl imidazolium-5-amino tetrazolate (BMIMAT), with promising energetic characteristics was synthesized by metathesis reaction. The synthesis involves two steps: in the first step, 10 mmol of the 5-amino tetrazole was added to 5 mmol of aqueous potassium carbonate solution. The solution was then concentrated by vacuum distillation to separate a white solid. The crude product obtained was recrystallized from a 1:4 water-alcohol mixture to get potassium-5-amino tetrazolate (K-AT) (Ernst et al., 2007; McCrary et al., 2012). The 1-butyl-3-methyl-imidazolium-5-amino tetrazolate (BMIMAT) was prepared by metathesis reaction of 1-butyl-3-methyl-imidazolium chloride (BMIMCl) and K-AT. Herein, 3 mmol of BMIMCl and 4.5 mmol of K-AT were added to 150 ml of acetone and stirred for 24 h. Insoluble KCl was removed as a precipitate and the filtrate was vacuum-dried to yield a yellow viscous liquid. The liquid was again dissolved in acetone, filtered and the solvent was removed by vacuum distillation. The synthetic scheme of BMIMAT is presented in **Fig. 3.2**.

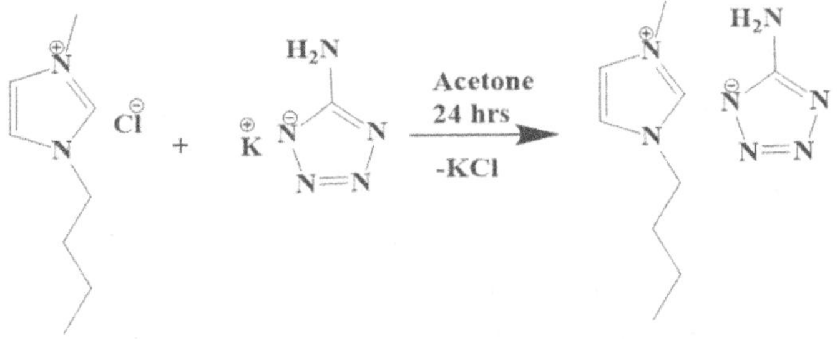

Figure 3.2. Synthetic scheme of BMIMAT.

3.2.5 Characterization

NMR analyses were done in deuteriated DMSO solvents with a BRUKER AVANCE 400 MHz (with respect to ^{1}H nucleus) FT-NMR spectrometer. Power-gated proton decoupling and inverse-gated proton decoupling methods were applied for ^{13}C NMR and ^{15}N NMR, respectively.

FTIR spectrum was recorded using Nicolet iS50 FTIR spectrometer. The spectrum was recorded over the wavenumber range of 4000-400 cm^{-1} at a resolution of 4 cm^{-1} by smearing the sample between two KBr cells to form a thin film.

Electrospray mass spectrometry was used to detect the cation and anion parts of ionic liquids/ Mass spectrometric determination was conducted using a Bruker Q-TOF (COMPACT) mass spectrometer. The solvent used was methanol-water-acetic acid (49:49:2) mixture. The ionization mode was set as positive for cation part determination and negative for anion part, with end plate offset: 1000V, capillary voltage: 6000 V, nebulizer: 0.4 bar, dry gas: 4L/min, dry temperature: 200°C. The MS/MS was carried out by setting the mass window as 2 amu and the collision energies in the range of 0-50 eV.

The decomposition products of the IL were determined using Py-GC-MS. Py-GC-MS consists of Thermo Electron trace Ultra GC directly coupled to a mass spectrometer and the SGE pyrolyzer. The capillary column in GC is 30 m, 0.25 mm ID (PDMS with 5% phenyl, SLB-5MS, Supelco). The ion source was set at 200 °C, and the column temperature program from 35-200 °C. The pyrolysis of IL was executed at 300 °C with a heating rate of 5 °C /min.

3.2.5.1 Thermal analysis

The thermal stability of IL was studied using a thermogravimetric analyzer (TA instrument-Q50) in the N_2 atmosphere. Glass transition temperature was determined using a differential scanning calorimeter (DSC Q20, TA instrument, USA) in the N_2 atmosphere. In order to measure the glass transition temperature, the sample was first cooled from +50°C to -75 °C and then heated from -75 °C to +50°C, at a rate of 10 °C/min. Kinetic analysis of thermal decomposition of the compounds was carried out by the method of Kissinger and Ozawa-Flynn-Wall

(OFW), using the differential heating rate method(Flynn and Wall, 1966; Kissinger, 1957; Ozawa, 1965). The peak values of exothermic decomposition from the DTA thermogram were used for the calculation of activation energy. Experiments were carried out from 50 to 300 °C at varying heating rates of 2, 4, 6, and 8 °C/min.

In order to establish a mathematical relation between heating rate, rate of conversion, and temperature it is essential to determine the activation energy, Ea, from thermal analysis data. The mathematical methods for thermal analysis can be model-fitted or model-free isoconversional (Williams et al., 2020). To provide a reliable kinetic mechanism, we combined model fitting and model-free approaches for kinetic evaluation studies. The activation energy E_a is attained from Eq (3.1) for the Kissinger method with multiple heating rate.

$$\ln\left(\frac{\beta}{T^2}\right) = \ln\left(\frac{AR}{T}\right) - \frac{Ea}{RT} \qquad (3.1)$$

Here β is the heating rate, and T, the temperature in Kelvin scale at a fixed conversion value of α. A is the pre-exponential factor, E_a is the activation energy at a given conversion and R is the gas constant. For the Kissinger method $\ln\left(\frac{\beta}{T^2}\right)$ was plotted against $\frac{1}{T}$ giving rise to a straight line whose slope yields activation energy. A model-free method developed by Ozawa, Flynn, and Wall, the activation energy is given by Eq. (3.2)

$$Ea = \frac{k*R}{b} \qquad (3.2)$$

where k is the slope obtained by plotting $\ln(\beta)$ vs $\frac{1}{T}$, assuming the initial value of b as 0.457.

3.3 Results and Discussion

3.3.1 Computational study of ionic liquids with BMIM Cation

The molecular structure of each ionic liquid is optimized using B3LYP/6-311+G(d,p) level of theory employing Gaussian 09 software. The optimized structure, HOMO-LUMO gap, binding energy and heat of formation of the five different ionic liquids are listed in **Table 3.1**. Even though the cation-anion interactions in ionic liquids are mainly coulombic in nature, there exists H-bonding interaction between the acidic H in the C2 position of the imidazolium ring with the anions (Dong et al., 2006). In the five ionic liquids, the cation tends to be coplanar with the anions over a strong H-bonding along the C2-H plane. The C2-H----anion bond lengths are 1.78, 1.73, 1.70, and 1.78 Å for NO_3^-, Tz, DATz and AT anions, respectively. For dinitramide anion, multiple interaction sites are observed with a C2-H bond length of 2.02 and 2.0 Å. Typically the H-bonding interaction between the cation and anion in ILs should be shorter than the sum of the van der Waals radii of H (1.20 Å) and N (1.55 Å) to form a maximum H---N bond distance of 2.75 Å (Xuan et al., 2011).

The properties of ILs are generally associated with their interaction energy, which is defined as the energy difference between IL (cation-anion pair) and the total energies of the cationic and anionic species. The large interaction energies of these ionic liquids compared to a hydrogen bond energy of 50 kJ/mol suggest the existence of strong electrostatic interaction of the cation with each anion (Thomas et al., 2019).

Table 3.1 Theoretically optimized structure, HOMO-LUMO gap, Heat of formation (HOF) and binding energy of BMIM cation with nitrate (N), dinitramide (DN), 5-amino-1,2,4-triazolate(Tz), 3,5-diamino-1,2,4-triazolate (DTz) and 5-amino tetrazolate (AT) anions.

IL	Optimized Structure	HOMO-LUMO gap (eV)	HOF (kJ/mol)	Binding energy(kJ/mol)
BMIMN	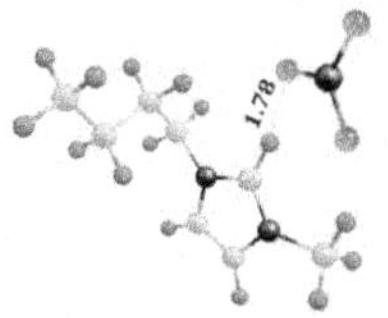	4.2	-166.1	379.54
BMIMDN	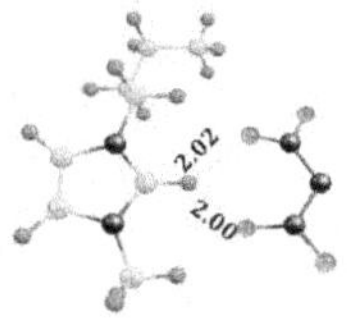	4.8	+65.81	405.04
BMIMTz	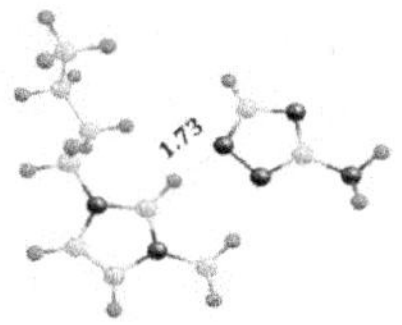	3.3	+219.5	241.21
BMIMDTz	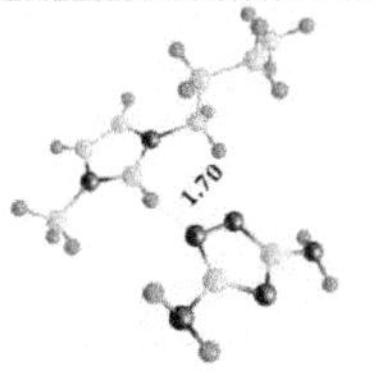	3.0	+218.4	185.75
BMIMAT	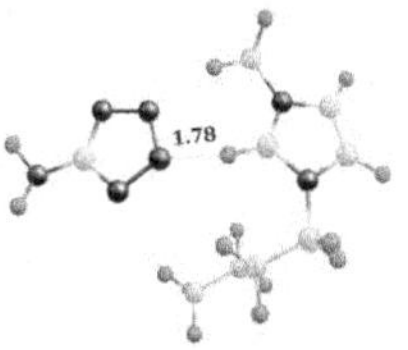	3.6	+317.2	301.13

The decrease in the computed cation-anion interaction energies are observed when moving from nitrate and dinitramide anions to heterocyclic aromatic anions. The relative stability of the five ionic liquids is in the order BMIMDN > BMIMN > BMIMAT >BMIMTz > BMIMDTz. Nitrate and dinitramide possess relatively localized negative charge, allowing for strong Coulombic attraction with the corresponding cations. Azolate anions display extensive electron delocalization across their conjugated heteroaromatic frameworks. Such delocalization lowers the effective charge density and thereby diminishes the electrostatic component of the cation–anion interaction. As a result, ion pairs containing triazolate and tetrazolate anions exhibit weaker binding energy compared to those formed with nitrate and dinitramide, consistent with their reduced interaction energies. This reduction in electrostatic strength is an intrinsic characteristic of azolate anions and has important implications for tuning the physicochemical properties of ionic-liquid systems (Panda et al., 2017).

The chemical stability and intrinsic reactivity of an IL can be effectively predicted by analyzing frontier molecular orbitals, particularly the energy separation between the highest occupied molecular orbital (HOMO) and the lowest unoccupied molecular orbital (LUMO). The HOMO-LUMO energy gap reflects the ease of electron excitation or charge transfer within the system. A larger energy gap corresponds to a stable IL with lower propensity toward chemical transformation. Conversely, a smaller HOMO–LUMO gap suggests increased susceptibility to chemical reactions (Ofem et al., 2023). Among the ionic liquids investigated, IL with triazolate anion is unstable and more reactive relative to other anions. The order of stability follows the same as that of interaction energy (Vastag et al., 2018), ie. BMIMDN >BMIMN >BMIMAT > BMIMTz > BMIMDTz with the energy gaps of 4.8 eV > 4.2 eV >3.6 eV >3.3 eV > 3.0 eV. The lower HOMO–LUMO energy gaps observed for the ionic liquids under study, compared to conventional liquids, imply their high reactivity and energetic characteristics. Thus, frontier orbital analysis provides a valuable predictive framework for tailoring ionic liquids with specific performance characteristics, whether the objective is to design a highly stable IL or to design one with enhanced energetic reactivity. The studies indicated that BMIMAT possesses favorable stability parameters compared to other azolate based ionic liquids.

The heat of formation (HOF), energy content of the material, is crucial in assessing the potential energetic performance. The gas phase HOF can be calculated by different levels of theory such as DFT-B3LYP, MP2, G2, and G3 (Glorian et al., 2021). The HOF of ILs is calculated by the G3 method encompassing the method of atomization (Thomas et al., 2015). The HOF increases as the number of nitrogen atoms in the backbone increases. This arises due to the presence of high-energy N-N and C-N bonds, which store significant chemical energy due to their bond dissociation features and the tendency to release stable nitrogen as a decomposition product with a bond dissociation energy of 954 kJ/mol(Jafari and Keshavarz, 2016). Hence ILs with greater N content exhibit elevated heat of formation, contributing to the energy density and performance. The HOF determined theoretically for nitrate, DN, Tz, DTz and AT anions with BMIM cation are -166.1, 65.81, 219.5, 218.4 and 317.2 kJ/mol, respectively. The HOF of BMIM cation gives a higher value with AT anion, which can be a superior candidate for energetic application.

3.3.2 Theoretical performance evaluation of HAN with IL fuels

The specific impulse, a rocket performance parameter, is the thrust generated per unit weight flow rate of propellant consumed. Achieving higher Isp values is desirable for long-duration missions. The highest Isp can be achieved by a propellant formulation with the highest combustion chamber temperature (Tc) and evolution of low molecular weight ($\dot{m}$) exhaust gases. The Isp, Tc and $\dot{m}$ is related as, Isp $\propto \sqrt{\frac{Tc}{\dot{m}}}$.

Hydroxylammonium nitrate (HAN) is an energetic monopropellant characterized by a positive oxygen balance of 33.33%. Hence HAN has sufficient oxygen to support complete combustion of additional fuel components. Formulating HAN with suitable fuel additives in stoichiometrically optimized proportions can significantly enhance the overall performance of the propellant system. Such tailored HAN-fuel mixtures improve energy release, combustion efficiency, and specific impulse, making them attractive for advanced propulsion applications. Also EILs with oxygen-deficient anions must be convoyed by a suitable oxidizer to achieve the best performance(Sebastiao et al., 2014). As reported by Berg and Rovey (2013), binary mixture of ionic liquid fuel and an oxidizing salt such as hydroxyl ammonium nitrate (HAN) can function as a high-performance monopropellant due to the better solubility of HAN in EILs. The theoretically determined heat of formation is used as an input to calculate the ballistic properties such as specific impulse and chamber temperature (Berg and Rovey, 2013).

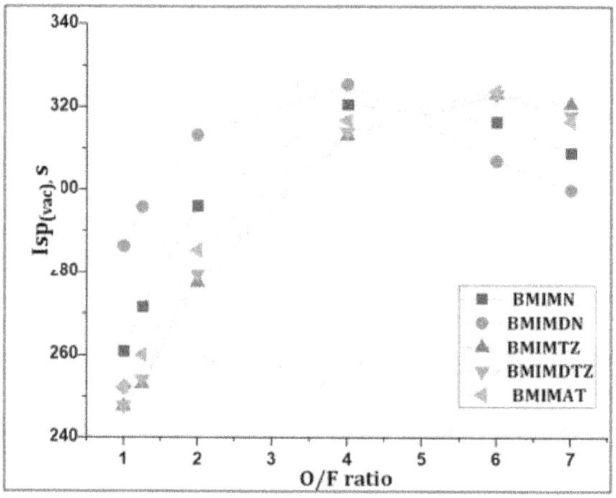

Figure 3.3 $Isp_{(vac)}$ vs O/F ratio of HAN-IL mix.

The HAN-IL formulation exhibits an improvement in specific impulse of nearly 90 seconds compared to neat 95 wt.% aqueous HAN solution ($Isp_{(vac)}$=235 s) with an augmented chamber temperature (Tc) of>2600 °C. **Fig. 3.3** shows that $Isp_{(vac)}$ increases with an increase in oxidizer to fuel ratio, attains a maximum value and decreases thereafter. Maximum Isp for nitrate and dinitramide-based ionic liquid fuels with HAN oxidizer is attained at an O/F of 4. Oxygen-rich IL provides higher Isp even at a lower O/F ratio. However, in the case of ILs with heteroatomic anions the maximum Isp was achieved with an oxidizer to fuel ratio in the range of 5-6. **Table 3.2** gives the $Isp_{(vac)}$ and corresponding chamber temperature for HAN fuel formulations at maximum O/F ratio.

Table 3.2 Maximum $Isp_{(vac)}$ and chamber temperature (Tc) for different HAN formulations.

Formulations	**Max Isp (s)**	**Tc (°C)**
HAN-BMIMN	320.7	2618
HAN-BMIMDN	325.49	2628
HAN-BMINATZ	323.07	2611
HAN-BMINDATZ	323.04	2599
HAN-BMIMAT	323.8	2600

The addition of a fuel component to HAN results in an enhancement of the performance which is directly related to chamber temperature, suggesting the competence of the ILs studied for energetic applications. Although all the ILs investigated in combination with HAN demonstrated improved performance, BMIMAT is selected for further detailed study owing to its high heat of formation value, and consequently the enhanced specific impulse (Isp). BMIMAT exhibits a relatively high positive HOF compared to other ILs due to the greater number of N-N bonds within the tetrazolate structure.

Theoretical studies reveal that an O/F ratio of 6 yields the maximum adiabatic flame temperature and favourable molecular weight of exhaust products, resulting in a peak in specific impulse. At lower O/F ratios, incomplete oxidation of the fuel leads to the formation of heavier intermediate species, thereby increasing the average molecular weight of the exhaust and reducing Isp. Conversely, in higher oxidizer ratio, lowering of the flame temperature and consequently lower performance is observed. **Fig. 3.4** shows the mole fractions of the major gaseous species evolved, as computed using NASA CEA, during the decomposition of the HAN–EIL system at the optimum O/F ratio. The major decomposition products of the BMIMAT–HAN system includes CO_2, N_2, and H_2O.

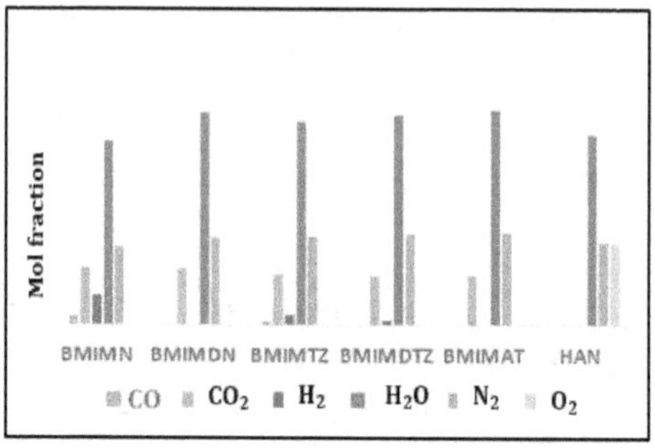

Figure 3.4 Mole fraction of gases evolved on decomposition of EIL-HAN mixture at maximum $Isp_{(vac)}$.

3.4 Characterisation of BMIMAT

5-amino tetrazole (AT), a nitrogen-rich compound, has found applications as a high-performance propellant and gas generator (Smiglak et al., 2010; Tao et al., 2008). AT and its derivatives behave as weak bases with inorganic acids and are hence often employed as cations in energetic salts. N-rich energetic ionic liquids containing AT as an anion can be realized by imparting an acidic nature to tetrazole by substitution at the C-5 position. Azolate anions have additional advantages of synthetic flexibility, electronic charge diffusion, and tailorable structural asymmetry for the building of a variety of ILs (Ogihara et al., 2004).Tetrazole possesses high energy density and produces environmentally benign N_2 gas as the combustion product. Xue et al., (2005) and Katritzky et al., (2006, 2005) have investigated energetic azolium-azolate ionic liquids with melting points falling below 100 °C. Tao et al. (2012) reported on the synthesis and stability of room-temperature ionic liquids with AT anion employing imidazolium cations with alkyl groups of different chain lengths in the 1 and 3 positions. Although literatures are available on the thermal characteristics of azole-based cations, the studies on tetrazoles as anions are inadequate.

Computation studies reveal that the main type of interaction between cation and anion in BMIMAT is the hydrogen bonding between the C2-H of cation and N of anion with a C2-H...N distance of 1.78Å. The anion tends to lie coplanar with the cation through a strongly directed and localized H-bonding with the acidic proton at the C2 position causing an asymmetric electron distribution in the rings (Noack et al., 2010). The topographical analysis of molecular electrostatic potential (MESP) plots is a tool to study the coordination of cation to anion (Kumar et al., 2014), which is electrostatic in nature. In the MESP plot the negative charge (blue region) is distributed spherically with the 5-amino tetrazolate anion and the positive charge (red region) is located around the imidazolium cation. **Fig. 3.5** gives the chemical structure and MESP of BMIMAT. The computed binding energy of -368.37 kJ/mol reveals dominant electrostatic interaction in BMIMAT as the hydrogen bonding contribution is generally limited to 50 kJ/mol (Hunt et al., 2015).

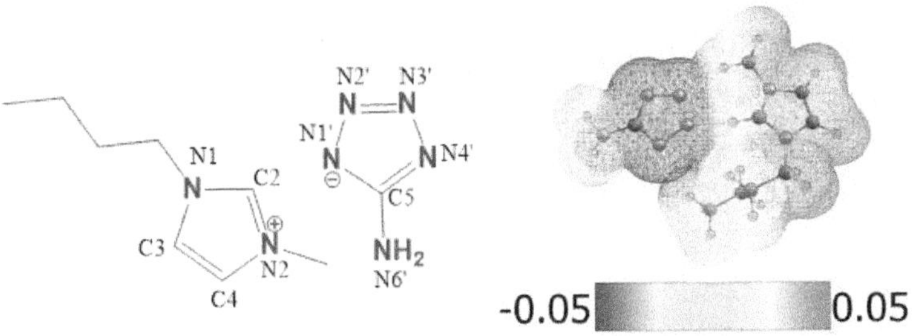

Figure 3.5 Chemical structure and MESP plots of BMIMAT (Color coding: -0.05 au. (Blue) to +0.05 au. (Red).

3.4.1 Spectroscopic analysis

The formation of BMIMAT is confirmed from the up-shielded resonance of the C2H proton at 9.2 ppm in the ^{1}H NMR spectrum when the anion changed from chloride to AT owing to the charge dispersion in the latter. Further, the ^{13}C signals of C2 in imidazolium cation is shifted from 139 ppm in BMIMCl to 136.7 ppm, and the C5 of tetrazole moiety is shifted from 158 ppm in 5-AT to 164 ppm with the formation of BMIMAT. ^{15}N NMR spectral data shows five resolved resonance signals from the five different types of N atoms. The signals at 183.3 ppm and 171.1 ppm arise from the resonance stabilized N atoms in the imidazolium ring, N1and N2 respectively. The signals at 377 ppm, and 277 ppm are attributed to N atoms of tetrazolate ring (N1′-N4′ and N2′-N3′). The peak corresponding to N atom of the amino group in tetrazole occurs at 36.9 ppm. NMR spectrum of BMIMAT is shown in **Fig. 3.6**.

^{1}H NMR: δ = 9.2(s, 1H), 7.8 (s, 1H), 7.7(s, 1H), 4.2(t, 2H, J=7.2 Hz), 3.8(s, 3H), 1.7-1.8(m, 2H), 1.2(sxt, 2H, J=7.4 Hz), 0.9(t, 3H, J=7.3 Hz) ppm.

^{13}C NMR: δ =164.2, 136.7, 123.8, 122.6, 48.8, 35.9, 31.6, 19.0, 13.6 ppm.

^{15}N NMR: δ =377.1, 277.2. 183.3. 171.1, 36.9 ppm.

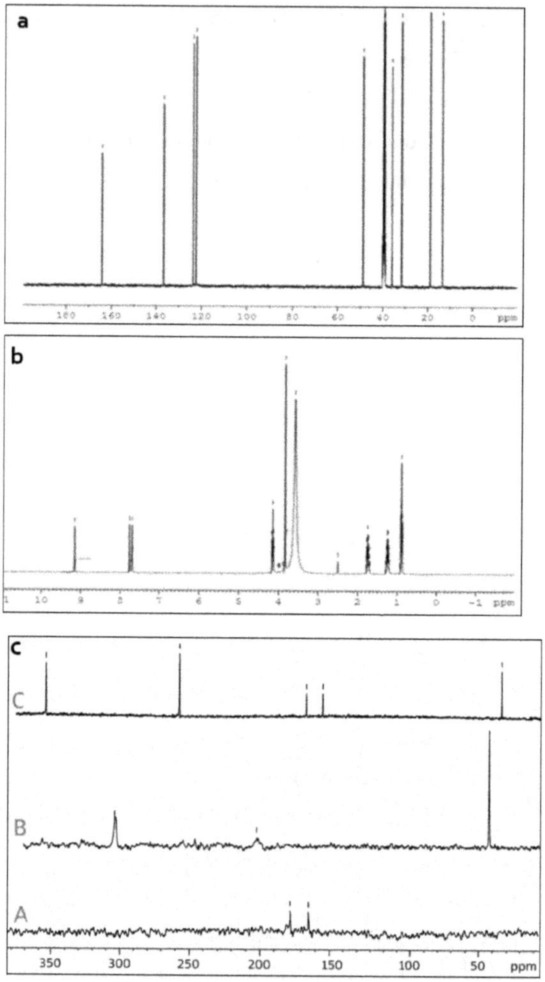

Figure 3.6 (a) ^{13}C NMR spectrum of BMIMAT, (b) ^{1}H NMR spectrum of BMIMAT and (c)^{15}N NMR spectrum of (A) BMIMCl, (B) 5-AT and (C) BMIMAT.

FTIR spectrum of BMIMAT with bands corresponding to imidazolium and tetrazolate moieties is presented in **Fig. 3.7.** The bands at 3157 cm^{-1} and 3118 cm^{-1} arise due to aromatic C-H vibrations of the imidazolium ring. The broad peak at 3450 cm^{-1} is due to the absorbed moisture. The characteristic peaks at 2962 cm^{-1} and 2879 cm^{-1} correspond to aliphatic C-H stretching vibrations. The peak at 1572 cm^{-1} corresponds to the ring C=C vibration of the imidazolium cation. The bands at 841 cm^{-1} and 752 cm^{-1} are related to imidazolium ring bending modes. The absorptions of tetrazole ring moiety are observed at 1466 cm^{-1} and 1516 cm^{-1}.

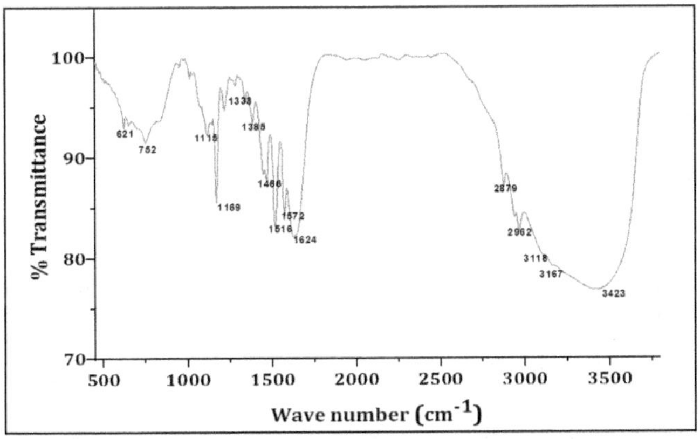

Figure 3.7 FTIR spectrum of BMIMAT.

ESI-MS is used to get structural information without fragmentation. **Fig. 3.8(a)** and **Fig. 38(b)** shows the positive ion mass spectrum of IL with m/z 139 and the negative ion mass spectrum with m/z 84, respectively. Other peaks in the spectrum due to solvents are also given in **Fig. 3.8(b)**. ESI-MS/MS analysis confirms the identity of an ion by isolating a specific precursor and analyzing its characteristic fragment ions. MS/MS spectrum of 84 m/z in **Fig. 3.8(c)** shows the fragmentation of anion, which confirm the structure of the anion species.

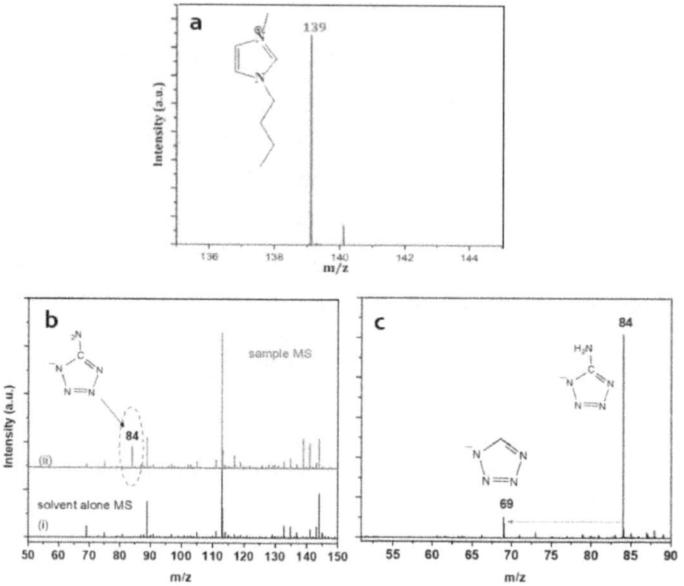

Figure 3.8 (a) Positive ion mass spectrum, (b.i) MS of the solvent, (b.ii) Negative ion mass spectrum of BMIMAT and (c) MS/MS spectrum of m/z 84.

3.4.2 Thermal Analysis and Kinetic Computation of BMIMAT

The thermal decomposition mechanism of ILs is very important for their real use in a wide array of applications. TG-MS studies are generally employed to comprehend the thermal decomposition mechanism of AT-based energetic material (Neutz et al., 2003; Piekiel and Zachariah, 2012). In this work, the kinetics and thermal decomposition mechanism of 1-butyl-3-methyl-imidazolium-5-amino tetrazolate (BMIMAT) is investigated by experimental and theoretical approaches.

The practical applications of ionic liquids are largely governed by their ability to remain in the liquid phase over a wide temperature range. Since ionic liquids do not exhibit a sharp melting point like conventional salts, the glass transition temperature (Tg) becomes a critical thermal parameter. Estimation of Tg provides valuable insight into the lower temperature limit at which the ionic liquid retains sufficient molecular mobility and fluidity for effective operation. The most usual way for experimentally determine Tg is differential scanning calorimetry where cooling of the room temperature ionic liquids (RTILs) may cause glass formation at low temperatures (Fredlake et al., 2004). Tg of BMIMAT determined by low-temperature DSC. The low-temperature DSC of BMIMAT shown in **Fig. 3.9(a).** Tg of BMIMAT is observed at -63°C, which relates to the structural features and interaction between the cation and anion. The lower value of Tg implies a weak interaction of ions due to their bulky and asymmetric nature. The thermal stability of BMIMAT is analyzed using TG. **Fig. 3.9(b)** gives the TG/DTG of BMIMAT. The peak decomposition temperature is 247 °C at a heating rate of 5 °C/min. The initial decomposition temperature in TG is assigned at 8% weight loss. The weight loss below 100 °C is attributed to the loss of moisture and solvent used for the synthesis, as confirmed by the pyrogram done at 100°C.

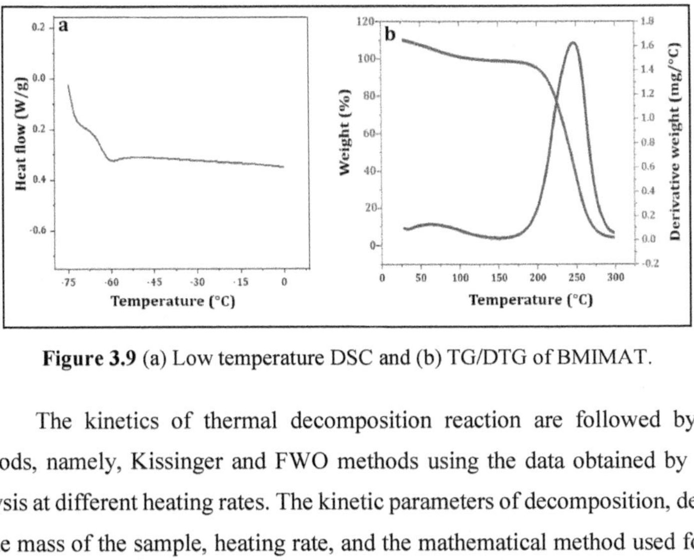

Figure 3.9 (a) Low temperature DSC and (b) TG/DTG of BMIMAT.

The kinetics of thermal decomposition reaction are followed by two methods, namely, Kissinger and FWO methods using the data obtained by TGA analysis at different heating rates. The kinetic parameters of decomposition, depend on the mass of the sample, heating rate, and the mathematical method used for the evaluation of kinetic data. It has been generally found that the model-free isoconversional method is preferred for multiple heating rates (Vyazovkin et al., 2011).The activation energy is calculated in the Kissinger method using Eq (3.1) by plotting $\frac{\ln \beta}{T^2}$vs $\frac{1000}{T}$, and the plot of ln β against $\frac{1}{T}$ will give the activation energy at different conversion levels for the model-free FWO method.

The experimental evaluation of E_a is done using the Kissinger method and isoconversional FWO method as shown in **Fig. 3.10** and **Fig. 3.11**, respectively.

The activation energy for thermal decomposition calculated using the Kissinger method is 122.84 kJ/mol. Activation energy calculated using the FWO method using conversions between 0.2 and 0.8 with correlation coefficients not less than 0.98 is 127.63 kJ/mol. From the plot of E_a *vs* α, it is evident that activation energy decreases with the extent of conversion. This implies a more stable intermediate is formed during the initial decomposition process and as the decomposition proceeds, the consecutive intermediates/transition states have less activation energy/stability. Beyond 50% conversion, E_a is more or less constant, indicative of a single transition state. This resembles a single-step decomposition mechanism for BMIMAT (Koga et al., 2023). The activation energy calculated using Kissinger method and isoconversional FWO method are presented in **Table 3.3**.

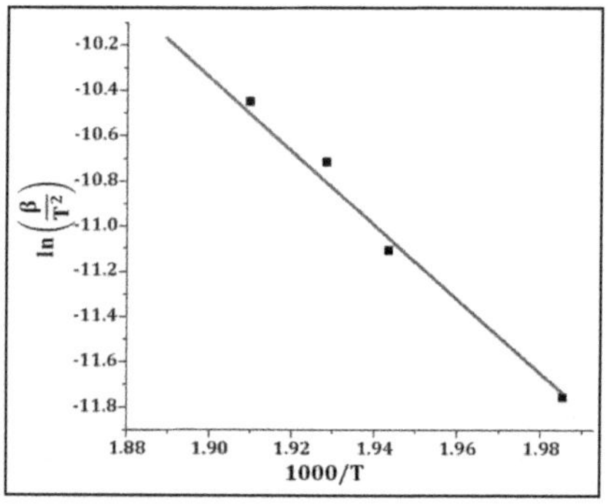

Figure 3.10 Kissinger plot of BMIMAT.

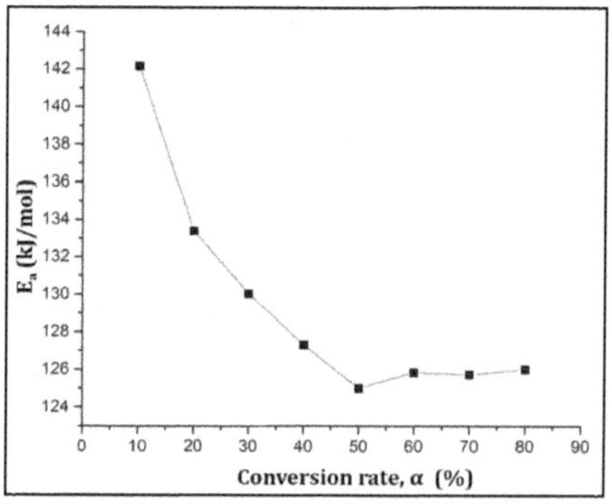

Figure 3.11 Plot of Activation energy (E_a) versus conversion (α) for the thermal decomposition of BMIMAT.

Table 3.3 Activation energy of thermal decomposition of BMIMAT calculated experimentally.

Method	Kissinger	FWO
E_a (kJ/mol)	122.84	127.63

3.4.3 Mechanism of thermal decomposition

To describe the possible ways of thermal decomposition, the ionic liquid (BMIMAT) is subjected to pyrolysis at 300 °C (Jiang et al., 2019) and the decomposition products separated by GC are identified by MS as shown in **Fig. 3.12**

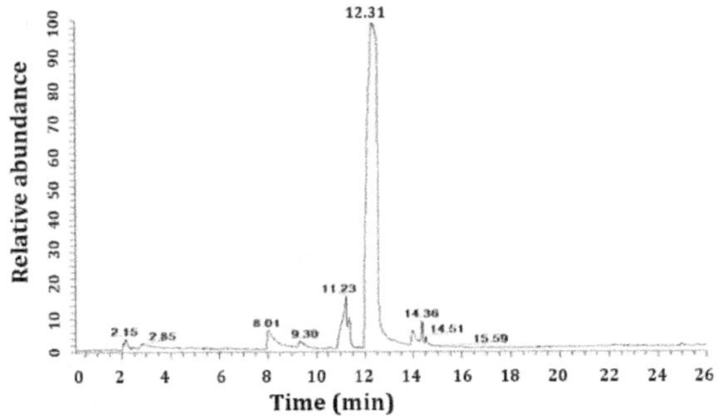

Retention time(min.)	8.01	11.23	12.31	13.95	14.36	14.51
Major Products	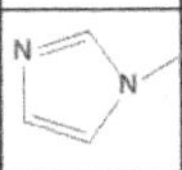	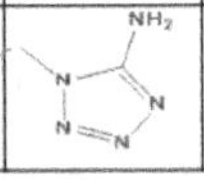	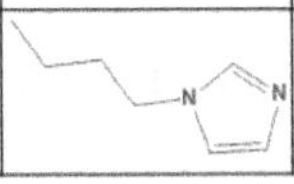	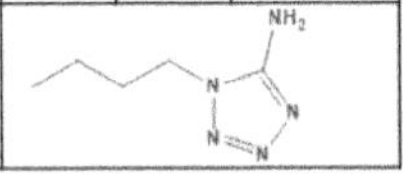		

Figure 3.12 Pyrogram of BMIMAT, pyrolyzed at 300°C.

The major peaks detected for gaseous decomposition products are 1-methyl-5-amino tetrazole (MAT, m/z=99), 1-butyl imidazole (BIM, m/z=124), 1-butyl-5-amino tetrazole (BAT, m/z=141) and 1-methylimidazole (MIM, m/z=82), suggesting the existence of S_N2 mechanism of decomposition. The dealkylation is initiated by the bimolecular nucleophilic attack of tetrazolate anion at the alkyl groups of imidazolium cation. The mechanism proposed for the decomposition based on the products attained on pyrolysis is schematically presented in **Fig. 3.13.**

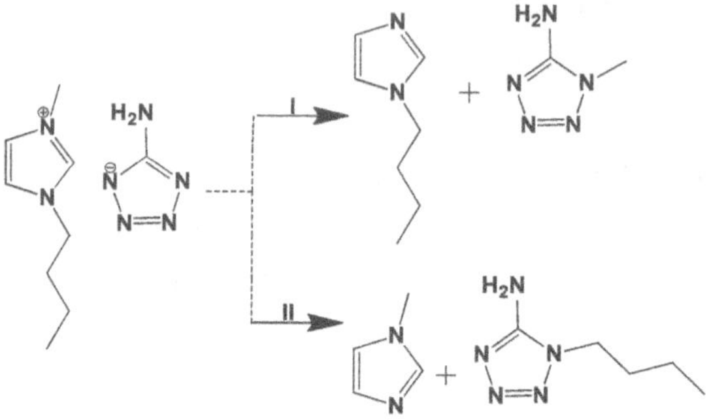

Figure 3.13 Schematic representation of thermal decomposition pathway of BMIMAT.

The decomposition of BMIMAT via S_N2 pathway is dominated by the direct nucleophilic attack of the anion on the alky side chain. The nucleophilic attack will preferentially occur on the methyl substituent, due to the higher steric hindrance of the longer butyl chain. This is evident from the higher concentration of 1-butyl imidazole than 1-methyl imidazole in the decomposition products. Moreover, a positive charge on the N1 nitrogen induces a positive charge on the methyl group which further facilitate the nucleophilic attack. (Maton et al., 2013).

3.4.4 Computational study on decomposition of BMIMAT

The S_N2 mechanism is further complimented by theoretical studies using B3LYP at 6-311+G(d,p) level of DFT. In Pathway I, the transition state is categorized by the breaking of the C-N bond at C...N distance of 1.472 Å and the formation of C-N bond distance of 1.444 Å forming 1-methyl-5-aminotetrazole. The breaking of the C-N bond at C...N distance of 1.482 Å and the formation of a C-N bond distance of 1.476 Å forming 1-butyl-5-aminotetrazole attributes the transition state for Pathway II. Demethylation in Pathway I, proceeds with an E_a of 145.6 kJ/mol and is identified as the favorable route of decomposition with the formation of 1-butylimidazole and 1-methyl-5-aminotetrazole. Debutylation proceeds via Pathway II, with a higher E_a of 153.89 kJ/mol. Due to the steric hindrance of tetrazolate anion, the S_N2 reaction has a higher probability to happen predominantly at the N3-CH_3 position. The higher concentration of 1-methyl-5-aminotetrazole than 1-butyl-5-aminotetrazole in the decomposition products indicates the preference of decomposition pathway I. Moreover, the major decomposition product identified in pyrogram,1-butylimidazole, further confirms the preference of decomposition pathway I. DFT studies further show that the thermal decomposition reaction is slightly exothermic by ~38 kJ/mol. **Table 3.4** provides the comparison of activation energy obtained by the experimental and computational studies. The activation energy values calculated via Kissinger and FWO methods are lower compared to that obtained from quantum chemical calculations. This may be due to the assumption of gas phase in the computational method and the lattice energy change associated with ionic liquids in the actual liquid phase (Koga et al., 2023).The energy profile and transition state structures are shown in **Fig. 3.14**.

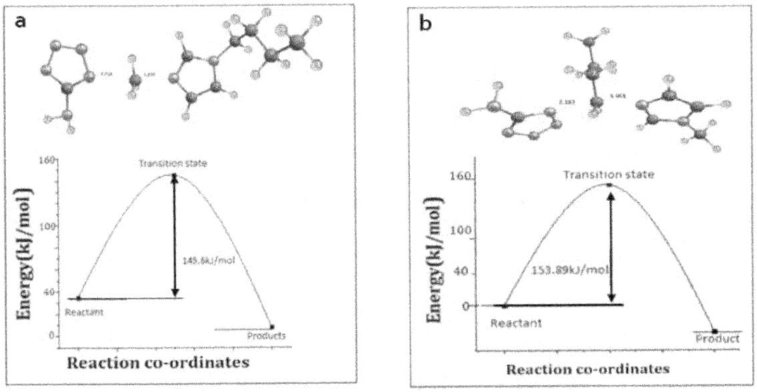

Figure 3.14 Transition states and energy profile diagram for (a) Pathway-I (b) Pathway-II for thermal decomposition of BMIMAT.

Table 3.4 Quantum chemical and experimental activation energy values.

Method	**Theoretical**		**Experimental**	
	Pathway I	Pathway II	Kissinger	FWO
Activation energy, E_a(kJ/mol)	145.63	153.89	122.84	127.63

The quantum chemical calculation of different ionic liquids yielding 1-butyl imidazole and methylated anion further validates the results (Kroon et al., 2007). The anion has a prominent impact on the decomposition temperature of ILs and the nucleophilic anion present lowers the decomposition temperature (Neutz et al., 2003). It is imperative to note that tetrazolate anion provides the lowest onset and peak decomposition temperatures indicating its advantage as an energetic fuel. The variation in onset decomposition temperature observed for ILs containing BMIM cation with different anions reported in the literature is compared with the present system in **Table 3.5.** The onset decomposition temperature is a key parameter for assessing thermal stability; however, the corresponding kinetic parameters do not necessarily follow the same trend. The calculation of kinetic parameters is undertaken to gain deeper insight into the decomposition kinetics of BMIMAT. In particular, the activation energy (E_a) provides an estimate of the minimum energy input required to initiate decomposition or combustion, which is critical for evaluating its suitability for specific applications. The activation energy values give an idea of how easily the corresponding activated complexes can be formed from the reactants. The smaller anions such as chlorides and nitrate have lower activation energy since the five coordinated transition complexes in S_N2 reactions are easily formed, compared to BF_4 and tetrazolate anions with higher steric factors.

Table 3.5 Variation of activation energy, onset decomposition temperature, and peak decomposition temperature of ILs with BMIM cation and different anions.

IL	**E_a (kJ/mol)**	**Tonset (°C)**	**Tpeak (°C)**	**Ref.**
BMIMAT	145.6	230	260	This work
BMIMCl	127	254	284	(Cao and Mu, 2014; Fredlake et al., 2004; Huddleston et al., 2001; Kroon et al., 2007)
$BMIMNO_3$	120.7	290	315	
$BMIMBF_4$	195	361	440	

3.5 Studies on HAN–BMIMAT system

Thermal stability studies at a heating rate of 5 °C/min indicate distinct decomposition behaviours for HAN and BMIMAT. HAN exhibits an onset decomposition temperature of 158 °C and a peak decomposition temperature of 195 °C. BMIMAT remains stable up to ~165 °C and exhibits a peak decomposition temperature of 247 °C. Monopropellant formulations are prepared by combining 80 wt.% HAN as the oxidizer with BMIMAT as the energetic fuel in an O/F ratio of 1, and the thermal behaviour is evaluated.

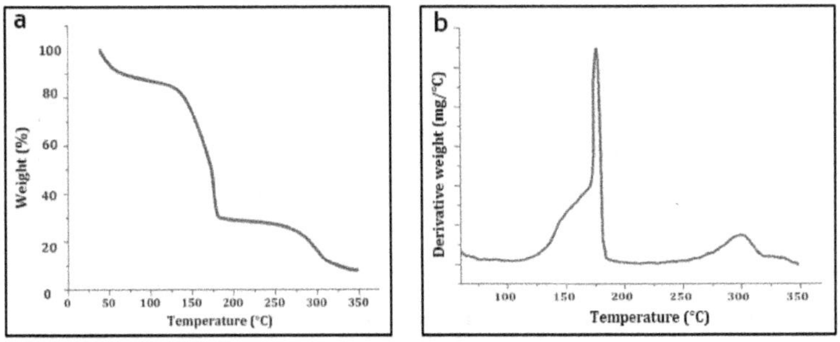

Figure 3.15 (a) TG and (b) DTG for HAN-BMIMAT at O/F: 1.

Thermal analysis of the HAN–BMIMAT (O/F:1) mixture reveals that the monopropellant formulation remains thermally stable up to 120 °C. Beyond this temperature, a distinct two-stage decomposition profile is observed. The TG and DTG of HAN-BMIMAT (O/F:1) mixture is shown in **Fig. 3.15(a)** and **Fig. 3.15(b),** respectively. Kinetic parameters for these decomposition events are determined using the Kissinger variable heating-rate method, which relies on the DTG peak maxima to calculate activation energies (E_a). The Kissinger plot for the first and second stage of decomposition of HAN-BMIMAT (1:1) formulation is shown in **Fig. 3.16**. The E_a values derived for the first and second decomposition stages are 103.4 kJ/mol and 166.7 kJ/mol, respectively. The peak decomposition temperatures and the corresponding E_a values are given in **Table 3.6.**

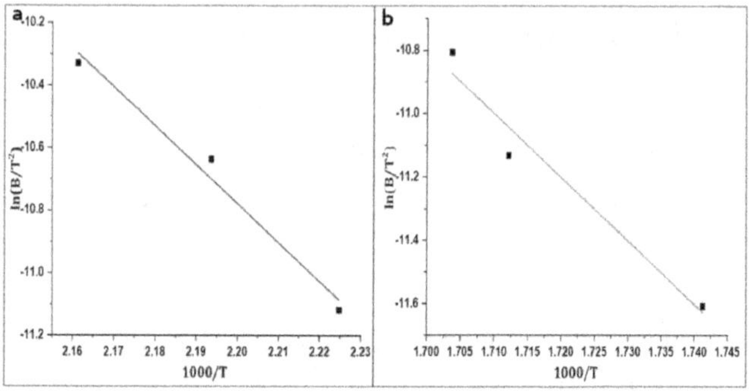

Figure 3.16 Kissinger plot for HAN-BMIMAT (O/F:1) formulation (a) first stage and (b) second stage of decomposition.

Table 3.6 E_a for the first and second decomposition stages of the HAN-BMIMAT at O/F:1.

Decomposition	**E_a (kJ/mol)**
Ist stage	103.4
IInd stage	166.7

The presence of two peaks in the HAN-BMIMAT thermal decomposition is attributed to the formation of double salts, specifically hydroxylammonium-5-amino tetrazolate and BMIMN. This interpretation is supported by the fact that the second-stage E_a value aligns closely with literature-reported E_a values for BMIMN decomposition (Liu et al., 2020). Computational analysis further corroborates the possibility of double-salt formation, supporting the proposed decomposition mechanism. The optimized geometry of HAN-BMIMAT mixture is presented in **Fig. 3.17**. The interionic distance, defined as the separation between the charge centres of the cation and anion, serves as an indicator of the strength of ionic interactions within the system. For the HA^+--AT^- ion pair, an interionic distance of 1.575 Å is observed. This is significantly shorter than the corresponding distances of 1.66Å observed for HA^+--NO_3^-, and the distance of 1.78Å observed between $BMIM^+$--AT^- ion pair. The comparatively reduced separation in the HA^+--AT^- system indicates a stronger electrostatic attraction and enhanced binding affinity between hydroxylammonium and 5-amino tetrazolate. Such strong cation–anion interactions support the likelihood of double-salt formation in the HAN-BMIMAT mixture, complementing the thermal and kinetic observations.

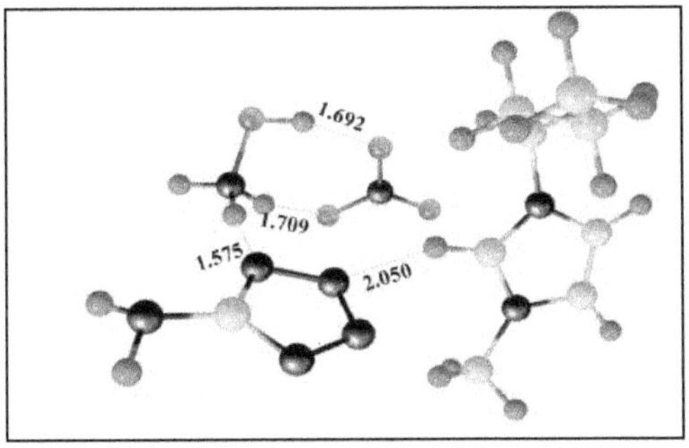

Figure 3.17 Optimized geometry of HAN-BMIMAT.

ESI-MS analysis of the HAN–BMIMAT mixture provides additional evidence for the formation of double salts (Wainwright et al., 2018). ESI-MS of the mixture for positive m/z measured for 50-400 m/z is given in **Fig. 3.18(a)**. In the positive-ion mode, a dominant peak at *m/z* 139 corresponds to the 1-butyl-3-methylimidazolium ($BMIM^+$) cation. ESI-MS of the mixture for negative m/z measured for a 50-500 m/z is give in **Fig. 3.18(b).** The signal at *m/z* 147 corresponds to the combination of [AT-H] with $[NO_3]^-$ (5-amino tertazole-Nitrate) which has a mass of 147 amu. A combination of $[BMIM]^+$ with $2[NO_3]^-$ has a mass of 263 amu and corresponds with the 263 m/z peak, consistent with the formation of a double salt. The peak at 113 m/z may be due to the combination of $[NH_2OH]$ $[AT]^-$ (mass 117 amu), and indicate pairing of hydroxyl amine molecule with 5-amino tetrazolate anion. This supports the coexistence of BMIM-nitrate and AT-based adducts, confirming double-salt formation in the HAN-BMIMAT system.

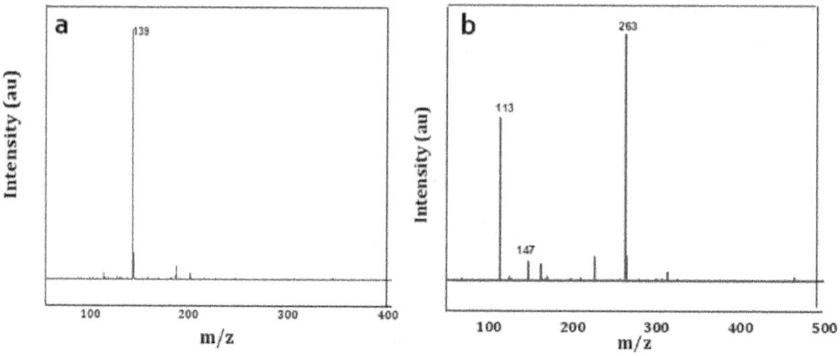

Figure 3.18 (a) Positive ion spectrum and (b) negative ion spectrum for HAN-BMIMAT.

The major gaseous products evolved during the thermal decomposition of the HAN-BMIMAT mixture are identified using TG-MS analysis, as shown in **Fig. 3.19**. The mass spectral data reveal distinct signals corresponding to *m/z* 30, attributed to nitric oxide (NO); *m/z* 28, for molecular nitrogen (N_2); *m/z* 44, associated with carbon dioxide (CO_2) or nitrous oxide (N_2O); and *m/z* 18, corresponding to water (H_2O). Additionally, a signal at *m/z* 85 is consistent with fragments originating from the 5-amino tetrazolate (AT). The collective presence of these species indicates a multistep decomposition pathway for the HAN-BMIMAT system.

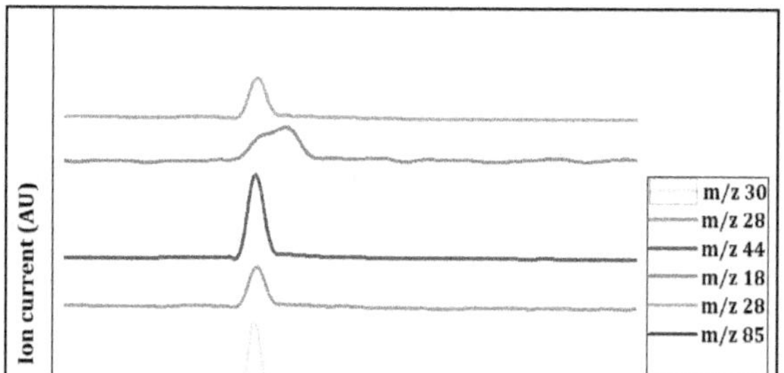

Figure 3.19 TG-MS spectrum of the decomposition products of HAN-BMIMAT mixture.

The stochiometric reaction of HAN and BMIMAT monopropellant system, assuming complete oxidation of fuel and formation of the thermodynamically favourable products, can be represented as

$$2C_9H_{17}N_7 + 27\ N_2H_4O_4 \rightarrow 18CO_2 + 61H_2O + 41N_2$$

The O/F ratio obtained from this ideal decomposition reaction is 5.8, which is very close to the theoretical O/F value of 6, identified as the optimum ratio for achieving maximum Isp in HAN-BMIMAT formulation.

Thermal analysis of the HAN-BMIMAT mixtures prepared at various O/F ratios is conducted to assess how composition influences the decomposition behavior. The DTG graph of HAN-BMIMAT mixtures of various composition is shown in **Fig. 3.20** and the corresponding onset and peak decomposition temperatures are presented in **Table 3.7.** At O/F ratio of 1 and 1.5, HAN-BMIMAT mixture shows two peak in the DTG. Above this, the DTG of HAN–BMIMAT mixtures show only one peak. Notably, the formulation corresponding to the theoretically optimized oxidizer-to-fuel ratio (O/F = 6), which delivers a predicted specific impulse ($Isp_{(vac)}$) of 323.8 s, exhibits a single step decomposition with an onset decomposition temperature of 135 °C. This onset temperature is lower than the individual onset temperatures of pure HAN and BMIMAT, indicating enhanced energetic characteristics arising from synergistic interactions between the oxidizer and the ionic-liquid fuel. The single-step decomposition observed can be due to the presence of oxidizer-rich composition. The O/F:6 composition not only maximizes theoretical performance but also presents a modified thermal response consistent with highly energetic monopropellant systems. The single step decomposition with low onset temperature, high specific impulse, low-toxic combustion products and broad liquid range makes the HAN-BMIMAT mixture a potential monopropellant for spacecraft propulsion.

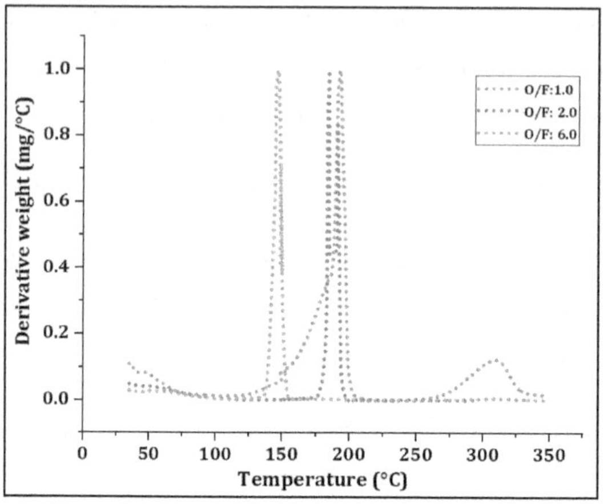

Figure 3.20 DTG of HAN-BMIMAT at different O/F ratios.

Table 3.7 Tonset and Tpeak decomposition temperatures of HAN-BMIMAT mixtures at various O/F ratios.

HAN-BMIMAT (O/F ratio)	Tonset (°C)	T peak (°C)
1.0	130	191 and 311
1.5	145	193 and 317
2.0	174	192
4.0	153	168
6.0	135	146

3.6 Conclusions

Energetic ionic liquids have been investigated as fuel for hydroxyl ammonium nitrate-based monopropellant system to improve the rocket performance parameter, specific impulse. Electronic structure and thermochemical parameters of ionic liquids with 1-butyl-3-methyl imidazolium (BMIM) cation and five energetic anions viz., nitrate(N), dinitramide (DN), 3-amino-1,2,4-triazolate (Tz), 3,5-diamino-1,2,4-triazolate (DTz) and 5-amino tetrazolate (AT) are determined by Density Functional theory (DFT). The theoretical performance evaluation of HAN with energetic ionic liquids (EIL) as fuel is performed with NASA Chemical Equilibrium with Applications. All HAN-EIL combinations have increased the specific impulse of the system by ~90 s at an optimum oxidizer to fuel ratio with augmented chamber temperature of >2500 °C in comparison to HAN alone.

5-amino tetrazolate anion imparts relatively high positive HOF compared to other anions, which is attributed to their greater number of N—N bonds. The energetic ionic liquid BMIMAT selected for further studies is synthesized and characterized. BMIMAT exhibited onset and peak decomposition temperatures of 165 and 247 °C, respectively. The kinetics of the decomposition reaction is studied by Kissinger and FWO methods. Based on the decomposition products determined by Py-GC-MS, a bimolecular nucleophilic substitution mechanism is proposed and supported by DFT analysis. The thermal decomposition of the HAN–BMIMAT mixture proceeds via a double decomposition mechanism, attributed to the strong interaction between the AT anion and the hydroxylammonium cation, as confirmed by DFT analysis. The HAN-BMIMAT formulation at O/F ratio of 6 exhibited a single stage decomposition with onset and peak decomposition temperatures of 135 °C and 146 °C, respectively. Single step decomposition with low onset temperature, high specific impulse, nontoxic combustion products and broad liquid range make the HAN–BMIMAT mixture a potential monopropellant for spacecraft propulsion.

CHAPTER 4

HYDROXYL AMMONIUM NITRATE – IMIDAZOLIUM CATION BASED MONOPROPELLANT: EFFECT OF SIDE CHAIN ON IONIC LIQUID ENERGETICS

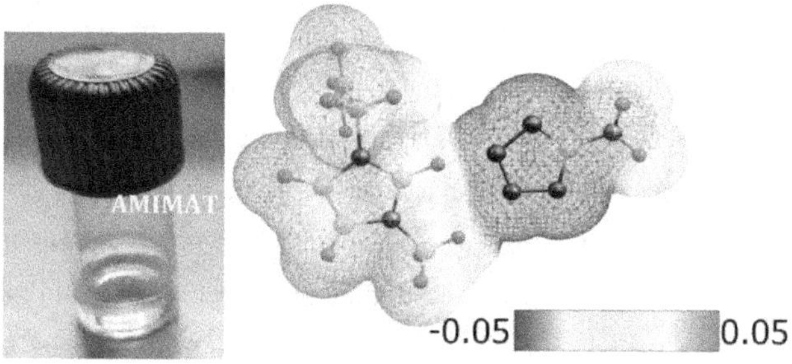

ChemistrySelect

Research Article
doi.org/10.1002/slct.202405705

Chemistry Europe
European Chemical Societies Publishing

www.chemistryselect.org

Design and Evaluation of 1-Allyl-3-Methyl Imidazolium-5-Amino Tetrazolate as Green Energetic Fuel

Sreelekshmi Karunakaran Radhika,[a, b] Raji Saraswathi,[b] Deepthi Thomas,[a] Vijayalakshmi Kunduchi Periya,*[a] and Prabhakaran Kuttan[b]

4.1 Introduction

Ionic liquids (IL) have been a topic of research over the last decades since the discovery of ethyl ammonium nitrate, the first IL made by Paul Walden in 1914. ILs are salts composed of organic cations and organic or inorganic anions with a low melting point. A wide variety of ILs can be designed by several suitable combinations of cations and anions and fine-tune their properties with different functionalities to suit specific applications. Due to these desirable and exceptional properties, ILs find diverse applications as green solvents (Greer et al., 2020; Wang et al., 2020), electrolytes (Asnawi et al., 2021), catalysts (Bartlewicz et al., 2020), lubricants (Verma et al., 2022), plasticizers (Miri et al., 2021) and as energetic materials (Zertal et al., 2022).

A key strategy in designing EILs is the selection of appropriate cations and anions that can synergistically enhance their energetic performance. Imidazolium-based cations, for instance, are widely studied due to their stable positive charge, excellent chemical and thermal stability, and the ability to modify their properties through side-chain substitutions (Kuznetsov, 2021; Schneider et al., 2008). The presence of an unsaturated side chain aids in enhancing the energetics, lowering of decomposition temperature, increasing the density and lowering of melting point (Hawker et al., 2014) compared to ionic liquids with the conventional alkyl chain. The density of ionic liquids containing unsaturated side chain substituents is higher due to an increase in packing efficiency. This also attributes to the lowering of viscosity due to fewer degrees of freedom of the ionic liquid. Moreover the unsaturated chain facilitates decomposition and hence the ILs will have lower decomposition temperatures. (Zhang and Shreeve, 2013).

In this context, the present work focuses on the design and evaluation of novel EILs based on 1-alkyl-3-methylimidazolium cation (alkyl=ethyl, butyl, allyl and vinyl) paired with five different anions: nitrate (N), dinitramide (DN), 5-amino-1,2,4-triazolate (Tz), 3,5-diamino-1,2,4-triazolate (DTz), and 5-amino tetrazolate (AT) anions. This study aims to explore the potential of substituted imidazolium cations paired with energetic anions as next-generation energetic materials for propellant applications.

4.2 Experimental

4.2.1 Materials

5-amino tetrazole (97 %), potassium carbonate (99 %) and 1-allyl-3-methylimidazolium chloride (≥98 %) were procured from Sigma-Aldrich, USA. The analytical reagent grade acetone used as solvent was procured from Merck India Ltd Mumbai. The chemicals were used without any further purification. Distilled water was used for the preparation of solutions.

4.2.2 Computational methodology

Density Functional Theory (DFT) calculations were conducted to determine the optimized molecular geometries of each ion pair using the Becke, three-parameter, Lee-Yang-Parr (B3LYP) hybrid functional at the 6-311+G(d,p) basis set. This level of theory provides a balance between computational efficiency and accuracy by incorporating diffuse and polarization functions in the basis set, which are particularly important for describing ionic interactions. The Gaussian 09 software package (Corrêa et al., 2008) was used for all calculations, ensuring consistency in computational methodology. To investigate the decomposition mechanism of AMIMAT, the same level of DFT was used. This method is widely used for studying organic and ionic systems due to its ability to model electronic correlation effects. The transition state structure was located and validated by confirming the presence of a single imaginary frequency in the vibrational analysis, which is a standard criterion for identifying transition states. Additionally, intrinsic reaction coordinate (IRC) calculations (Maeda et al., 2015) were performed to trace the minimum energy pathway ensuring that the identified transition state correctly connects the reactant and product structures. The heat of formation for individual ionic liquids was determined using the Gaussian-3 MP2 (G3MP2) *ab initio* composite method (Thomas et al., 2015). This high-level quantum chemical method systematically accounts for electron correlation effects and provides highly accurate thermochemical data, making it well-suited for calculating heat of formation (HOF). The choice of G3MP2 ensures a reliable estimation of reaction energetics by refining electronic structure calculations through a combination of MP2 and coupled-cluster corrections. The wavefunction data obtained from the optimisation results were utilised for the AIM analysis using the multiwfn program, to determine the critical point. The points in space where$\nabla \rho(r) = 0$, termed critical points. The energetic characteristics of nitrate, DN, Tz, DTz, and AT anions with cations having different side chain substitutions such as ethyl (1-ethyl-3-methylimidazolium – EMIM), butyl (1-butyl-3-methylimidazolium – BMIM), allyl (1-allyl-3-methylimidazolium – AMIM) and vinyl (1-vinyl-3-methylimidazolium – VMIM) groups were studied. Chemical structure of the cations and anions are presented in **Figure 4.1**.

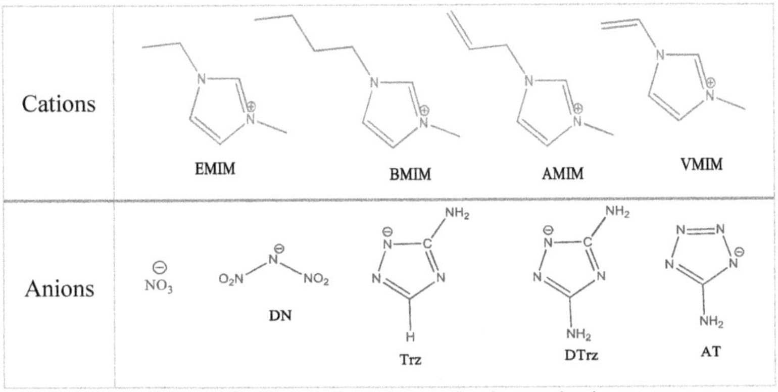

Figure 4.1. Chemical structures of cations and anions.

4.2.3 Theoretical performance evaluation

The theoretical prediction of the propellant performance was carried out by equilibrium combustion analysis using NASA Chemical Equilibrium with Applications (CEA) computer code (Frem, 2018). Mixtures of EILs (fuel) with oxidizer have been investigated in different oxidizer-to-fuel ratio under equilibrium conditions to determine the optimal high-performance monopropellant composition. The theoretical performance parameters including $Isp_{(vac)}$ and combustion temperature at various O/F weight ratios (O/F=1 to 7) were evaluated, assuming a chamber pressure of 20 MPa, a nozzle expansion to atmospheric pressure and the area ratio, A_e/A_t=50. Additional inputs to the model were the heat of formation (determined theoretically) and reactant's temperature of 298 K.

4.2.4 Synthesis of 1-allyl-3-methyl imidazolium-5-amino tetrazolate

The ionic liquid, 1-allyl-3-methylimidazolium-5-amino tetrazolate (AMIMAT), reported here for the first time was synthesized in two steps: In the first step, 10 mmol of the 5-amino tetrazole (5-AT) was added to 5 mmol of aqueous potassium carbonate solution. The solution was then concentrated by vacuum distillation to separate a white solid. The crude product obtained was recrystallized from 1:4 water-alcohol mixture to get potassium-5-amino tetrazolate (K-AT). In the second step, AMIMAT was prepared by metathesis reaction of 2 mmol 1-allyl-3-methylimidazolium chloride (AMIMCl) and 3.5 mmol of K-AT in acetone. Insoluble KCl was removed as a precipitate and the filtrate was vacuum-dried. The synthetic scheme of AMIMAT is presented in **Fig. 4.2**. After drying, the AMIMAT was obtained as a yellow liquid at room temperature.

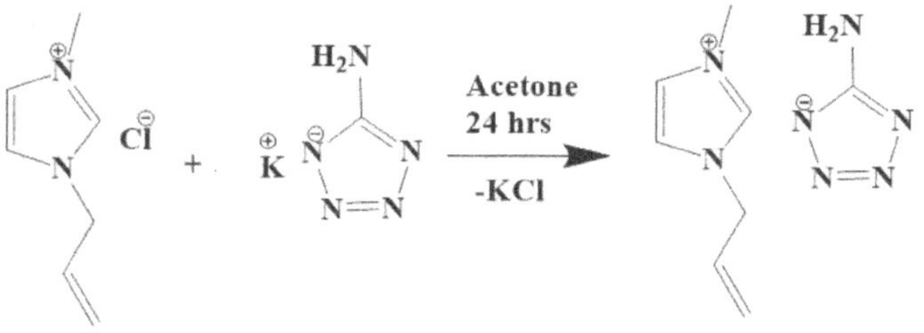

Figure 4.2 The synthetic scheme of AMIMAT.

4.2.5 Characterization

BRUKER AVANCE FT-NMR spectrometer was used for NMR analysis. For ^{13}C NMR and ^{15}N NMR, power-gated proton decoupling and inverse-gated proton decoupling methods were applied. NMR analyses were carried out in deuteriated DMSO solvent with 400 MHz for ^{1}H nucleus.

Fourier transform infrared analysis was performed using Nicolet iS50 FTIR spectrometer in the wave number range from 4000-400 cm^{-1} with a resolution of 4 cm^{-1}. The spectra were recorded by smearing the sample between two KBr cells to form a thin film.

Electrospray mass spectrometry was used to detect the cation and anion parts of ionic liquids. The sample for electrospray mass spectroscopy was prepared by diluting the ionic liquid with a methanol-water-acetic acid mixture (49:49:2). The ionization mode was set as positive for cation part determination and negative for anion part, with End plate offset: 1000V, capillary voltage: 6000 V, nebulizer: 0.4 bar, dry gas: 4L/min, dry temperature: 200°C. Mass spectrometric determination was conducted using a Bruker Q-TOF (COMPACT) mass spectrometer.

Thermogravimetric analysis (TGA) was performed using TA-Q50 instrument to determine the thermal stability of IL in the N_2 atmosphere from 50 to 400°C.

Pyrolysis-GC-MS was carried out to identify the decomposition products of the IL under study. Py-GC-MS consist of Thermo Electron trace Ultra GC coupled to a Thermo Electron Polaris Q (Quadrupole ion trap) mass spectrometer (Thermo Electron Corporation, USA) and the SGE pyrolyzer (Pyrojector II, Australia). The pyrolysis of AMIMAT was carried out at 250°C.

4.2.5.1 Thermal analysis

The thermal stability and kinetic analysis of the sample were carried out by Thermogravimetric analysis (TGA) under non-isothermal conditions. Kinetic analysis of thermal decomposition of the compound was performed by Model-free isoconversional methods for the evaluation of activation energy (E_a). The thermal analysis was performed at heating rates of 2, 4, 6, and 8°C/min in N_2 atmosphere. The thermogram can be converted into conversion rate (α) vs temperature relation for different heating rates in increments of 0.1 in the range of $0.3 < \alpha < 0.8$. The activation energy, E_a at respective α can be calculated by Kissinger-Akahira-Sunose (KAS)(Kissinger, 1957) and Ozawa-Flynn-Wall (OFW)(Flynn and Wall, 1966; Ozawa, 1965) methods.

In the KAS method, the activation energy E_a is obtained from Eq (4.1) with multiple heating rate methods.

$$\ln\left(\frac{\beta}{T^2}\right) = \ln\left(\frac{AR}{Ea.g(x)}\right) - \frac{Ea}{RT} \qquad (4.1)$$

Here β is the heating rate, T is the temperature in Kelvin scale,and $g(x)$is the reaction model integral conversion function. A is the pre-exponential factor, E_a is the activation energy at a given conversion and R is the gas constant. Ea is obtained from the slope of the linear plot of $\ln\left(\frac{\beta}{T^2}\right)$ vs $1000/T$ at different heating rates. The slope of the straight line equals to E_a/R.

In FWO, the activation energy is given by Eq. (4.2)

$$Ea = (k.R)/b \qquad (4.2)$$

where the slope, k is obtained by plotting $\ln(\beta)$ vs $1000/T$ at different heating rates and assuming the initial value of **b** as 0.462.

The activation energy E_a is also determined for the model fitted Kissinger method by Eq (4.3) with multiple heating rate.

$$\ln\left(\frac{\beta}{T^2}\right) = \ln\left(\frac{AR}{T}\right) - \frac{Ea}{RT} \qquad (4.3)$$

Here β is the heating rate, and T, the temperature in Kelvin scale. A is the pre-exponential factor, E_a is the activation energy at a given conversion and R is the gas constant. For the Kissinger method, $\ln\left(\frac{\beta}{T^2}\right)$ was plotted against $\frac{1}{T}$ giving rise to a straight line whose slope yields activation energy, E_a.

4.3 Results and discussion

4.3.1 Theoretical study of ionic liquids

The design and synthesis of ionic liquids with energetic characteristics opens an avenue in the development of new and green fuels for propellant applications (Zhang et al., 2011). The combination of imidazolium cation - which is known for ease of synthesis, stability, and fluid properties - with energetic anions offers the choice of a high-performing low-toxic energetic fuel. The properties of

IL can be readily tuned by functionalities and choice of suitable side chain in the imidazolium cation (Sun et al., 2017). The optimized structure of ILs with AT anion and different substituted imidazolium cations is given in **Fig. 4.3.** In all ionic liquids, the cation tends to be coplanar with the anions over a strong H-bonding interaction along the C2-H plane. The properties of ILs are generally associated with their interaction energy, which is defined as the energy difference between IL (cation-anion pair) and the total energies of the cationic and anionic species. The large interaction energies of these ionic liquids compared to a hydrogen bond energy of 50 kJ/mol suggest the existence of strong electrostatic interaction of the cation with each anion(Thomas et al., 2019). For each ionic liquid with different anions and cations, the contribution of the fundamental forces will vary, hence interionic distance cannot be correlated with binding energy.

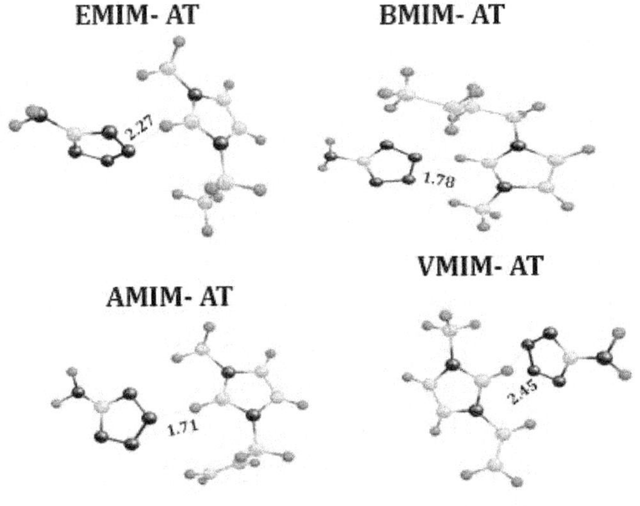

Figure 4.3 The optimized structures of EILs with AT anion.

The energetic performance of a material is strongly influenced by its heat of formation (HOF), which reflects the intrinsic energy content stored within its molecular framework. For ionic liquids intended for propellant applications, a positive HOF is particularly desirable, as it contributes directly to enhanced combustion and performance. The calculated HOF values for 1-alkyl-3-

methylimidazolium cations (alkyl = ethyl, butyl, allyl, and vinyl) paired with five anions viz., nitrate (N), dinitramide (DN), 5-amino-1,2,4-triazolate (Tz), 3,5-diamino-1,2,4-triazolate (DTz), and 5-amino tetrazolate (AT) are given in **Table 4.2.** An examination of the influence of side-chain substituents on the imidazolium ring paired with various anions reveals that an alkenyl substituents exhibit higher HOF values than their alkyl-substituted counterparts. This becomes more pronounced when the cations are combined with nitrogen-rich anions, leading to synergistic increases in HOF.

The ionic liquids incorporating heterocyclic anions exhibit significantly higher HOFs compared to those containing nitrate or dinitramide. The heterocyclic anions play a crucial role in tailoring the energetic characteristics of imidazolium ionic liquids for advanced propellant applications. This enhancement arises from the abundant high-energy N–N and C–N linkages present in triazolate and tetrazolate ring. Among these, ionic liquids incorporating tetrazolate anions exhibit the greatest increase in HOF, attributable to the higher number of N–N bonds within the tetrazole framework.

The predicted HOF of 1-vinyl-3-methylimidazolium cation with nitrate, DN, Tz, DTz, and AT anions range from 42 to 547 kJ/mol while for 1-allyl-3-methylimidazolium cation it is found between 6 to 485 kJ/mol. 5-amino tetrazolate anion is persuasive in imparting more energetic characteristics when compared to other anions in combination with VMIM cation. The higher HOFs observed for the vinyl-substituted cation suggest that the presence of the vinyl group contributes additional unsaturation and electronic delocalization, which can enhance the enthalpy. This underscores the combined role of both cationic and anionic components in modulating the energetic characteristics of imidazolium-based ionic liquids and highlights the importance of structural tuning when designing advanced energetic formulations.

The chemical stability or reactivity of the IL can be predicted by analysing the frontier orbital energy difference or the HOMO–LUMO gap. A large energy gap (eV) implies higher stability and vice versa. HOMO-LUMO analysis serves as a guide for the design of ILs with desired characteristics of high stability or increased reactivity (Ofem et al., 2023). The lower energy gap of ionic liquids under study than the conventional ionic liquids implies their high reactivity and energetic characteristics. The HOMO-LUMO gap and binding energy calculated are also presented in **Table 4.1**

Among the five ionic liquids investigated with AT anion, VMIMAT with the highest heat of formation exhibits a low energy gap of 2.8 eV implying high sensitivity and low thermal stability leading to challenges in handling, storage and safety. Hence AMIMAT having a slightly lower heat of formation with a higher HOMO-LUMO gap is considered for further synthesis and experimental evaluation.

Table 4.1 Theoretically calculated Heat of formation (HOF), HOMO-LUMO gap, and binding energy of cations with nitrate (N), dinitramide (DN), 5-amino-1,2,4-triazolate(Tz), 3,5-diamino-1,2,4-triazolate (DTz) and 5-amino tetrazolate (AT) anions.

Cation	Anion	HOF (kJ/mol)	HOMO-LUMO (eV)	BE (kJ/mol)
	NO_3^-	-96.5	4.2	377.81
	DN	+110.6	4.6	393.97
EMIM	ATZ	+290.6	3.2	243.71
	DATZ	+285.8	2.9	187.67
	AT	+375.4	3.8	287.06
	NO_3^-	-166.1	4.2	379.54
	DN	+65.81	4.8	405.04
BMIM	ATZ	+219.5	3.3	241.21
	DATZ	+218.4	3.0	185.75
	AT	+317.2	3.6	301.13
	NO_3^-	+6.13	4.2	375.48
	DN	+213.6	4.9	390.36
AMIM	ATZ	+379.5	3.1	238.16
	DATZ	+ 366.2	3.0	187.22
	AT	+485.0	3.6	291.46
	NO_3^-	+42.56	3.6	338.83
	DN	+254.9	4.2	348.53
VMIM	ATZ	+425.0	2.5	205.55
	DATZ	+420.5	2.2	150.43
	AT	+547.2	2.8	252.05

The allyl substituted, AMIM cation gives a higher value of HOF along with better stability with the energetic anions. Among the five ionic liquids investigated with AMIM cation, IL with triazolate anion is unstable and more reactive. The stability follows the order same as that of interaction energy, AMIMDN > AMIMN > AMIMAT > AMIMTz > AMIMDTz with the energy gaps of 4.9 eV > 4.2 eV >3.6 eV > 3.1>3.0 eV. The high HOMO–LUMO energy gap of AMIMN compared to AMIMAT implies better thermal stability for the former. But for monopropellant applications, the very high thermal stability will not support the purpose of a fuel. Also, the low energy gap of AMIMATz and AMIMDATz implies inferior thermal stability and may be difficult to handle due to high reactivity. AMIMAT having a moderate energy gap will have better thermal stability than ILs with nitrate and triazolate anion and therefore can be suggested as the most appropriate fuel candidate for propellant applications based on theoretical investigation. Hence, computational designing and evaluation of the electronic properties of ILs allows us to determine the best candidate fuel for monopropellant application

4.3.2 Theoretical evaluation of HAN-alkyl/alkenyl substituted imidazolium-5-amino tetrazolate monopropellant formulations

EILs with oxygen-deficient anions must be convoyed by a suitable oxidizer to achieve the best performance (Sebastiao et al., 2014). As reported by Berg and Rovey(2013), binary mixture of ionic liquid fuel and an oxidizing salt such as hydroxyl ammonium nitrate (HAN) can function as a high-performance monopropellant due to the better solubility of HAN in EILs. HAN and alkyl/alkenyl substituted imidazolium-5-amino tetrazolate were evaluated for their rocket performance parameters at different oxidizer to fuel ratio (O/F=1 to 7). The theoretically determined heat of formation was used as an input to calculate the ballistic properties such as specific impulse ($Isp_{(vac)}$) and chamber temperature.

Fig. 4.4 illustrates that the specific impulse ($Isp_{(vac)}$) increases progressively with an increase in the oxidiser-to-fuel (O/F) ratio, reaching a maximum at an O/F ratio of 5. This trend reflects improved oxidiser availability, which enhances the completeness of combustion and increases the energy release of the HAN–EIL monopropellant system. Beyond this optimum ratio, further addition of oxidiser is not beneficial, as it leads to oxidiser-rich conditions that dilute the energetic contribution of the fuel and limit performance gains. At a maximum O/F ratio of 5, $Isp_{(vac)}$ obtained for EMIMAT, BMIMAT, AMIMAT and VMIMAT are 324 s, 322 s, 324 s and 325 s, respectively.

The use of energetic ionic liquids (EILs) as fuels in HAN-based monopropellant formulations results in a noticeable enhancement in specific impulse, with performance being strongly influenced by the nature of the side-chain substituent on the imidazolium cation. For HAN–EIL formulations containing the AT anion at an O/F ratio of 1, the calculated $Isp_{(vac)}$ values for EMIMAT, BMIMAT, AMIMAT, and VMIMAT are 258 s, 252 s, 260 s, and 267 s, respectively. The observed variation in performance highlights the role of side-chain structure in modulating the energetic characteristics of the fuel. In particular, unsaturated side chains contribute to higher heats of formation and improved combustion efficiency, thereby imparting enhanced energetics to the overall monopropellant system.

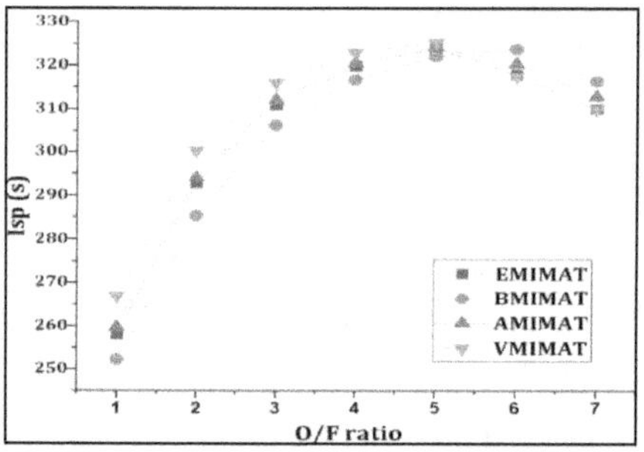

Figure 4.4 $Isp_{(vac)}$ for different O/F ratio of HAN-IL mix.

The HAN–1-allyl-3-methylimidazolium-5-amino tetrazolate (AMIMAT) formulation demonstrated a significant enhancement in performance, delivering a specific impulse of approximately 324 s, in contrast to the 235 s obtained for 95% HAN. This substantial improvement underscores the effectiveness of AMIMAT in augmenting the energetic output of HAN-based monopropellant systems. Based on its favorable theoretical performance predictions and energetic parameters,

AMIMAT is identified as the most suitable fuel component for HAN. Consequently, AMIMAT is synthesised and subjected to comprehensive physicochemical and structural characterization to validate its suitability for monopropellant applications. **Table 4.2** shown the comparion of computed electronic properties of substituted imidazolium ionic liquids with AT anion and theoretical maximum $Isp_{(vac)}$ in HAN formulation.

Table 4.2 Comparison of ILs for the binding energy, band gap, HOF, and Isp.

Parameters	**EMIMAT**	**BMIMAT**	**AMIMAT**	**VMIMAT**
Binding energy (kJ/mol)	287.06	301.13	291.46	252.05
HOMO-LUMO gap (eV)	3.8	3.6	3.6	2.8
Heat of formation (kJ/mol)	+375.4	+317.2	+485.0	+547.2
$Isp_{(vac)}$ (s) at max O/F ratio	324	322	324	325

4.4 Characterisation of AMIMAT

4.4.1 Spectroscopic analysis

AMIMAT is synthesized by metathesis reaction and characterized by different analytical tools such as NMR, FTIR, and ESI-MS to confirm the structure and purity.

^{1}H NMR:δ= 9.11 (s, 1H), 7.68 (s, 1H), 7.66 (s, 1H), 6.09 (m, 1H), 5.34 (m, 2H), 4.80 (d, 2H), 3.83 (s, 3H) ppm.

^{13}C NMR: δ= 163.7, 136.8, 131.9 , 123.8, 122.3, 120.5, 51, 40, 36 ppm.

^{15}N NMR: δ= 381.7, 282, 180.9, 172.1 36.8 ppm.

In the ^{1}H NMR spectra of AMIMAT, C2-H proton have a slight upfield resonance at 9.11 ppm from 9.48 ppm when the electron-withdrawing point charged chloride ion (AMIMCl) is replaced by resonance stabilized tetrazolate ion. The ^{13}C NMR spectra of the AMIMAT showed downfield resonance from 158 ppm to 164 ppm for C5 atom in the tetrazolate ring.The downfield shift was attributed to the resonance stabilized ring aromaticity of the tetrazolate anion (lower electron density at C5) in comparison with 5-amino tetrazole. The ^{1}H NMR and ^{13}C NMR spectra of AMIMAT are shown in **Fig. 4.5.**

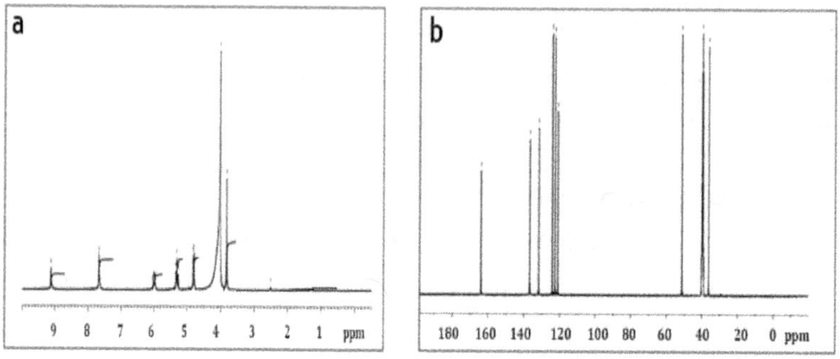

Figure 4.5 (a) ^{1}H NMR and (b) ^{13}C NMR spectrum of AMIMAT.

^{15}N NMR spectra exhibit five resonance signals resolved from the five different types of N atoms The signals at 180.9 and 172.1 ppm arise from the resonance stabilized N atoms in the imidazolium cation, N1and N2, respectively. The signals at 381.7 and 282 ppm are attributed to ring N atoms of tetrazolate anion (N1′, N4′ and N2′, N3′). The signals of N2′ and N3′ in tetrazolate anion appear upfield than N1′ and N4′ which are bonded to less electronegative carbon atoms. The peak corresponding to the N atom of the amino group in anion occurs at 36.8 ppm. ^{15}N NMR spectrum of AMIMAT is shown in **Fig. 4.6**

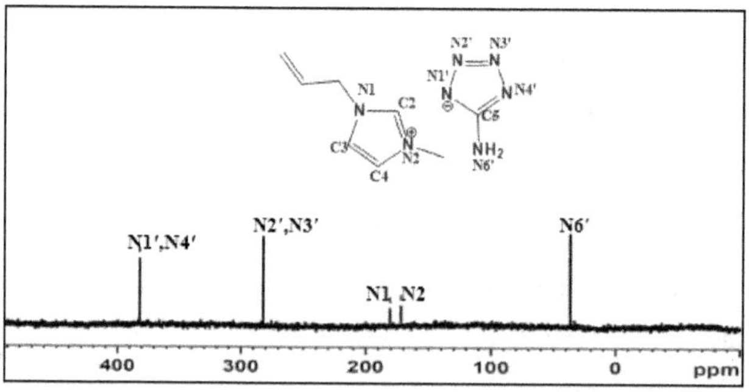

Figure 4.6 ^{15}N NMR spectrum of AMIMAT.

The synthesized ILs is also characterized by electrospray ionization mass spectrometry (ESI-MS) to further confirm its formation and purity. The ESI-MS of AMIMAT in positive and negative ion modes corresponds to the isolated cation and anion respectively. The mass-to-charge ratios detected are assigned to 1-allyl-3-methylimidazolium cation (m/z =123) and 5-amino tetrazolate anion (m/z =84). The ESI-MS spectrum of AMIMAT is presented in **Fig. 4.7**.

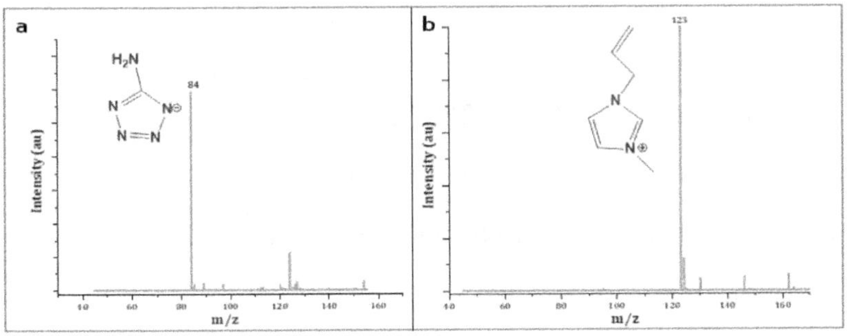

Figure 4.7 ESI-MS spectrum of (a) anionic and (b) cationic species.

FTIR spectrum of AMIMAT with bands corresponding to imidazolium and tetrazolate moieties are observed as shown in **Fig. 4.8**. The peaks at 3421 cm^{-1} and 3319 cm^{-1} corresponds to N-H stretching vibrations of the amine group in tetrazolate ring.The characteristic peaks at 3147 cm^{-1} and 3076 cm^{-1} arise due to asymmetrical and symmetrical C-H vibrations of the imidazolium ring. The peaks at 2958 cm^{-1} and 2873 cm^{-1} correspond to aliphatic C-H vibrations of alkyl chains attached to the imidazolium ring. The peaks at 1571, 1464 and1169 cm^{-1} correspond to ring C=C, C=N and N=N vibrations, respectively. The peaks at 1621 cm^{-1} and 1495 cm^{-1} are assigned to allyl C-C stretches.

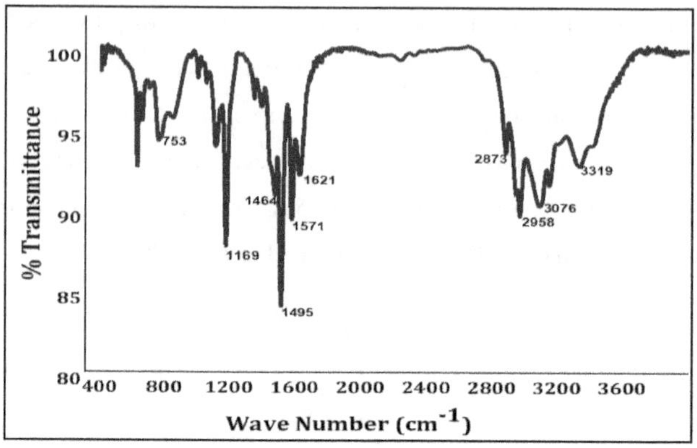

Figure 4.8 FTIR spectrum of AMIMAT.

4.4.2 Thermal analysis and Kinetic Computation of AMIMAT

The decomposition temperature of IL decides its operating temperature range.The thermal stability of AMIMAT is determined using thermogravimetry. The TGA curve exhibits a single step of decomposition with a peak decomposition temperature of 245 °C at a heating rate of 5 °C /min in N_2 atmosphere. The TGA graph of AMIMAT is shown in **Fig. 4.9.**

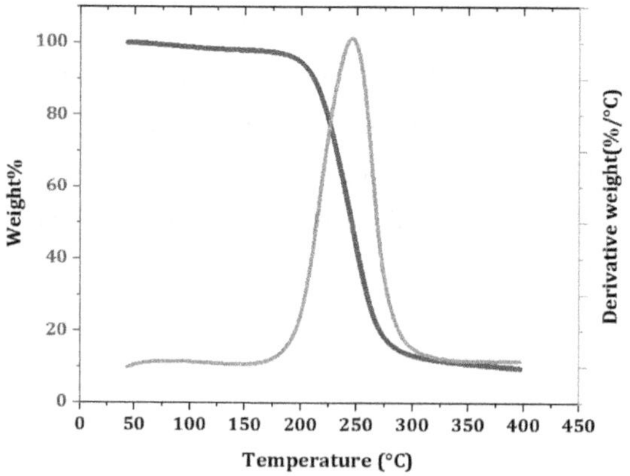

Figure 4.9 TG/DTG of AMIMAT.

The thermal stability of ILs depends on the anion coordination, hydrophilicity which is a measure of H-bonding strength, and nucleophilicity

(Maton et al., 2013). The onset (Tonset) and peak (Tpeak) decomposition temperatures of ILs with AMIM cations and different anions reported in the literature are summarized in **Table 4.3**. The peak decomposition temperature is observed in a broad range from 256 to 435 °C for ILs with different anions and the same cation. The relatively lower onset and peak decomposition temperatures observed for AMIMAT support the energetic nature pertinent to propellant formulations.

Table 4.3 Thermal analysis data of ILs with AMIM cation and different anions from literature.

ANION	Tonset (°C)	T peak (°C)	References
Azide	150	-	(Schneider et al., 2008)
Bis(trifluorometh anesulfonyl)imide	-	340	(Wang et al., 2013)
Chloride	249	270	(Hao et al., 2010; Cao and Mu, 2014)
Bromide	282	319	(Wang et al., 2007; Cao and Mu, 2014)
Tetrafluoroborate	400	-	(Min, et al., 2006)
Dicyanamide	278	480	(Zhang and Shreeve, 2011; Liang)
5-AT	180	256	This work (at 10 °C/min)

The kinetics of thermal decomposition are studied using the KAS and FWO methods by the differential heating rate method in thermogravimetric analysis. Model-free isoconversional methods (KAS and FWO) provide a more reliable kinetic analysis for the decomposition process at multiple heating rates. For the calculation of E_a values, conversion values between 0.1 and 0.8 having correlation coefficients not less than 0.98 are taken into account. The plot of temperature as a function of conversion rate (α) at different heating rates is shown in **Fig. 4.10(a)**. The activation energy for the decomposition of AMIMAT by the KAS method is calculated to be121.24 kJ/mol using Eq. (4.1).

In the FWO method, E_a values are determined using Eq. (4.2), and the average value is calculated to be 121.64 kJ/mol. The plot of variation in activation energy vs conversion rate for KAS and FWO methods is shown in **Fig. 4.10(b)**. The activation energy calculated by the KAS and FWO method remains more or less constant from a conversion rate of 10% to a high conversion rate of 80%, indicating the same decomposition mechanism throughout the processs (Zhang, 2020). The slight decrease in E_a values with the advance of conversion (above 40%) can be due to the secondary decomposition of primary decomposition products at higher temperatures.

The plot of $\ln\left(\frac{\beta}{T^2}\right)$ vs 1/T for the determination of E_a by the Kissinger method is given in **Fig. 4.10(c)**. The E_a obtained from the slope of the graph is 146.3 kJ/mol. The difference between this value and the activation energies calculated using isoconversional methods arises from the fundamental difference in the assumptions of the techniques. The Kissinger method evaluates E_a at the peak decomposition temperature, assuming a single-step reaction with a maximum reaction rate, whereas isoconversional methods determine E_a at different degrees of conversion over a range of heating rates, accounting for the multistep nature of the decomposition process. It has been generally considered that the model-free isoconversional method are preferred over the model fitted method for multiple heating rates to provide a reliable kinetic mechanism (Shakor et al., 2025).

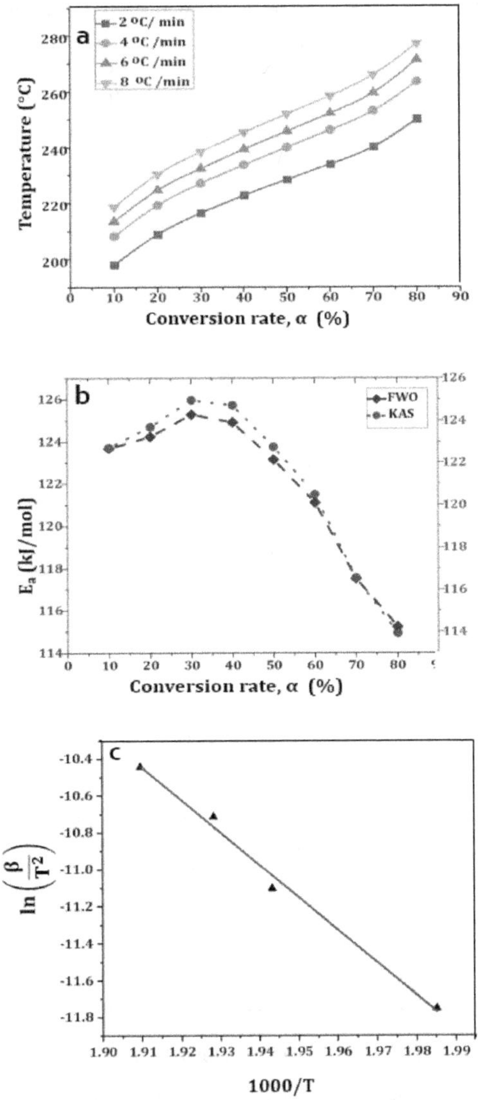

Figure 4.10 (a) Degree of conversion as a function of temperature at different heating rates (b) Activation energy (E_a) for different conversion percentages of the decomposition reaction calculated for KAS and FWO methods, and (c) Kissinger plot ($ln\left(\frac{\beta}{T^2}\right)$ vs 1/T) of thermal decomposition of AMIMAT.

4.4.3 Mechanism of thermal decomposition

Py-GC-MS analysis is carried out to identify the decomposition products of AMIMAT to predict the possible pathways of thermal decomposition. Pyrolysis is carried out at 250 °C and the decomposition products separated by GC are identified by MS. The pyrogram of AMIMAT is shown in **Fig. 4.11**.

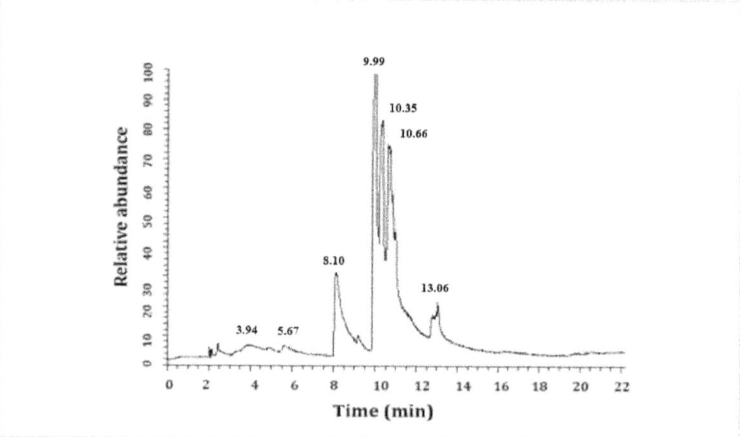

Retention time(min.)	3.94	5.67	8.10, 9.03	9.99, 10.35, 10.66	13.06
Major Products	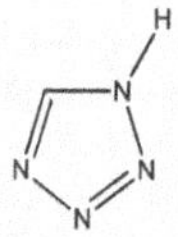	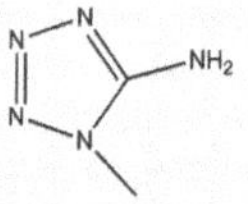	Methyl imidazole (isomers)	Allyl imidazole (isomers)	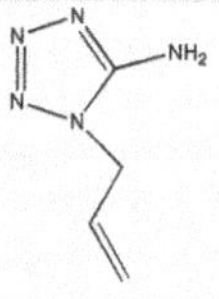

Figure 4.11 Pyrogram of AMIMAT sample at 250 °C.

The proposed decomposition mechanism based on the products identified on pyrolysis is schematically presented in **Fig. 4.12**. The major gaseous decomposition products detected are 1-allyl imidazole (AIM, m/z=108), 1-methyl imidazole (MIM, m/z=82), 1-methyl-5-amino tetrazole (MAT, m/z=99) and 1-allyl-5-amino tetrazole (AAT, m/z=125). The dealkylation and deallylation leading to the formation of 1-allyl imidazole and 1-methyl imidazole respectively, is due to the bimolecular nucleophilic attack (S_N2) of 5-amino tetrazolate anion at the alkyl/allyl group of imidazolium cation (Pathway I & II). The higher concentration of 1-allyl-5-amino tetrazole than 1-methyl-5-amino tetrazole in the decomposition products, shows the preference of decomposition pathway II.

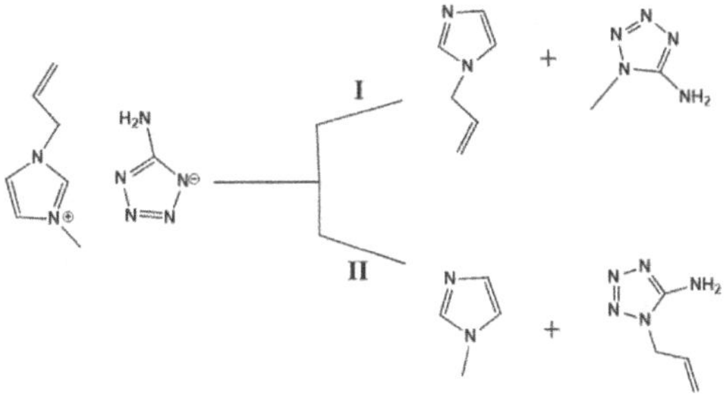

Figure 4.12 Schematic of the decomposition mechanism of AMIMAT.

4.4.4 Computational study on decomposition of AMIMAT

Computation studies reveal that the main type of interaction between cation and anion in AMIMAT is the hydrogen bonding between the C2-H of cation and N of anion with a C2-H...N distance of 1.71Å. The topographical analysis of the molecular electrostatic potential (MESP) plot given in **Fig. 4.13** shows the distribution of negative charge (blue region) spherically with the 5-amino tetrazolate anion and the positive charge (red region) is located around the imidazolium cation.

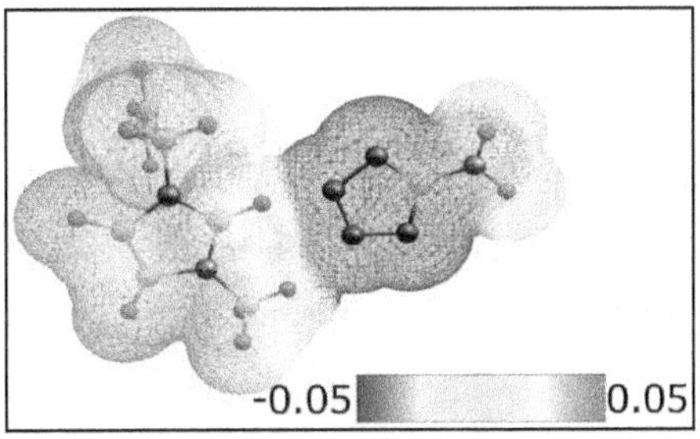

Figure 4.13 MESP plots of AMIMAT (Color coding: -0.05 au. (Blue) to +0.05 au. (Red)).

The decomposition reaction is analyzed by the computational method using B3LYP at 6-311+G(d,p) level of DFT. The transition state (TS) structures are identified and the activation energy is calculated. The TS-1 for bimolecular nucleophilic substitution reaction for Pathway I corresponding to demethylation is characterized by the breaking of the C-N bond at a C---N distance of 1.86Å and the formation of 1-methyl-5 amino tetrazole at a C-N bond distance of 2.12 Å with a computed E_a of 140.9 kJ/mol. The TS-2 for Pathway II is deallylation by the breaking of the C-N bond at a C---N distance of 1.94 Å and the formation of 1-allyl-5 amino tetrazole at a C-N bond distance of 2.21 Å with a computed E_a of 137.3 kJ/mol. Hence Py-GC-MS results agree well with the mechanism predicted by theoretical studies. The calculated activation energy is slightly lower for deallylation due to the presence of a reactive double bond. DFT studies show that the decomposition reaction is exothermic with an enthalpy of ~-42 kJ/mol. The energy profile and TS for the thermal decomposition by bimolecular nucleophilic substitution reaction (Pathway I and Pathway II) are shown in **Fig. 4.14**.

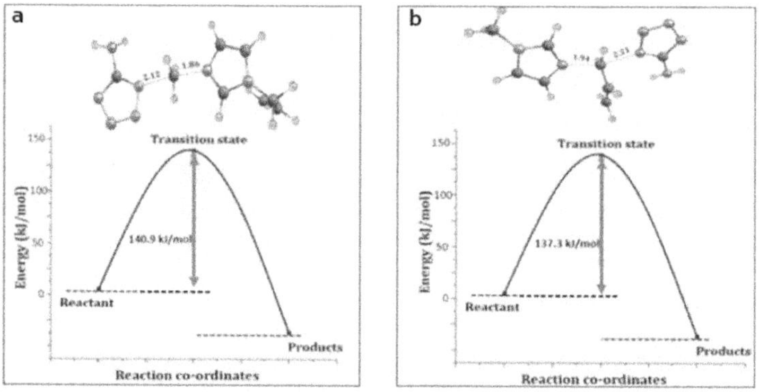

Figure 4.14 Energy profile and transition states of thermal decomposition reaction of AMIMAT (a) Pathway-I and (b) Pathway-II. Bond lengths in Å.

E_a values reveal the ease with which an activated complex can be formed from the corresponding reactants. **Table 4.4** gives a comparison of the activation energy values obtained by experimental isoconversional and computational methods. The differences in the experimental and theoretical values of E_a are due to the gas phase assumption in computational methods as compared to the actual liquid phase for the ionic liquids leading to changes in lattice energy (Thomas et al., 2016).

Table 4.4 Quantum chemical and experimental activation energy values.

	Computational study		Experimental Method		
E_a(kJ/mol)	Pathway I	Pathway II	KAS	FWO	Kissinger
	140.9	137.3	121.24	121.64	146.30

4.5 Studies on HAN-AMIMAT system

The stochiometric reaction of HAN and AMIMAT monopropellant system, assuming complete oxidation of fuel and formation of the thermodynamically favourable products, can be represented as:

$$4\ C_8H_{13}N_7 + 45\ N_2H_4O_4 \rightarrow 32\ CO_2 + 116\ H_2O + 59\ N_2$$

The O/F ratio obtained from this ideal decomposition reaction is 5.2, which is very close to the theoretical O/F value of 5, identified as the optimum ratio for achieving maximum Isp in HAN-AMIMAT formulation.

Monopropellant formulations are prepared by combining 80 wt% hydroxylammonium nitrate (HAN) with AMIMAT as the energetic fuel at different oxidizer-to-fuel (O/F) ratios. Formulations with O/F ratios of 1, 4, 5, and 7 are prepared, and their thermal behaviour is investigated. At lower O/F ratios, precipitation of the mixture is observed. This behaviour can be due to the formation of a double salt resulting from the interaction between hydroxylammonium nitrate and AMIMAT, leading to the formation of hydroxylammonium-5-amino tetrazolate and 1-allyl-3-methylimidazolium nitrate. For an O/F ratio of 1, precipitation occurred within 15–30 minutes. However, a delayed precipitation is observed in HAN-AMIMAT mixture at O/F ratios in the range of 2 to 4. On the other hand, at an O/F ratio higher than 4, the formulations remained clear and homogeneous with no visible phase separation. This improved stability is attributed to the enhanced solubility of the double-salt species in the higher water content associated with increased oxidiser loading, which promotes complete dissolution. **Fig. 4.15(a)** shows the photograph of HAN-AMIMAT formulations with at O/F ratio of 1, 4 and 5. The optical images of precipitate particles formed at O/F ratios of 1 and 4 are presented in **Fig. 4.15(b)**.

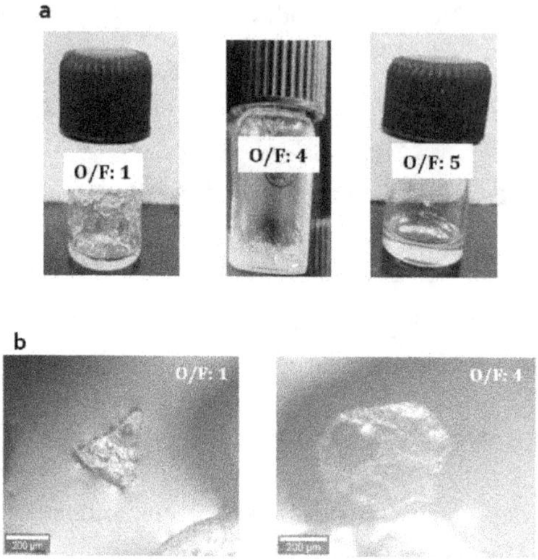

Figure 4.15 (a) Photograph of HAN-AMIMAT mix at O/F ratios of 1, 4 and 5 (b) Optical image of precipitate particles obtained for O/F ratio 1 and 4.

Theoretical studies are performed to substantiate the formation of a double salt in the HAN–AMIMAT system, which predicts favourable energetics and stability for the intermediate double-salt structures. The optimized geometry of HAN-AMIMAT mixture is presented in **Fig. 4.16**. The interionic distance between the charge centres of the cation and anion serves as an indicator of the strength of ionic interactions within the system. For the HA^+–AT^- ion pair, an interionic distance of 1.556 Å is observed. This is significantly shorter than the corresponding distances of 1.66 Å observed for HA^+–NO_3^-, and the distance of 1.71 Å observed between $AMIM^+$–AT^- ion pair. The comparatively reduced separation in the HA^+–AT^- system indicates a stronger electrostatic attraction and enhanced binding affinity between hydroxylammonium and amino tetrazolate. The presence of strong cation–anion interactions enhances the propensity for double-salt formation in the HAN–AMIMAT mixture, which manifests as the observed precipitate.

In the present work electron density at the bond critical points (BCP) are calculated and examined to understand the nature and strength of intermolecular interactions (Lu and Chen, 2012). The critical points (CP) are classified into four types viz, nuclear CPs, NCP(3, −3), bond CPs, BCP(3, −1), ring CPs, RCP(3, +1), and cage CPs, CCP(3, +3). The presence of a (3, -1) critical point and the bond path connecting two nuclei that passes through this point is considered for determining bond strength and nature of bonding interactions. The bond critical point connecting the N of 5-amino tetrazolate anion and NH_3OH^+ has an electron density of 0.07418 au which is higher than the value of 0.02139 au observed for the interaction between the $AMIM^+$ cation and AT^- anion. Furthermore, in HAN-AMIMAT system, AMIM…AT interaction is weakened, as evidenced by the reduction in the BCP electron density from 0.0524 au (observed for isolated AMIMAT) to lower values upon interaction with HAN. Since higher BCP electron density values correlate with stronger bonding interactions, these results clearly indicate stronger interactions between HA^+ and AT^-, which favor the formation of a double-salt structure. This strong ion pairing reduces the solubility of the resulting complex at lower oxidizer-to-fuel ratios, leading to the observed precipitation.

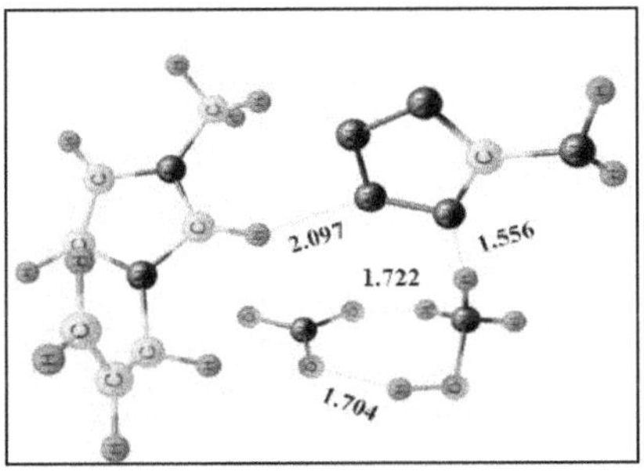

Figure 4.16 Computationally optimized HAN-AMIMAT geometry.

Thermal analysis of the HAN–AMIMAT mixtures prepared at various O/F ratios is conducted to assess how composition influences the decomposition behavior. The TG and DTG graph of HAN–AMIMAT mixtures of various composition is shown in **Fig. 4.17(a)** and **Fig. 4.17(b)**, respectively. The corresponding onset and peak decomposition temperatures obtained are presented in **Table 4.5**. It is observed that the onset and peak decomposition temperature of the formulation decrease with an increase in oxidizer content. The formulation corresponding to the theoretically optimized O/F ratio (O/F:5), which delivers a predicted specific impulse ($Isp_{(vac)}$) of 323.8 s, exhibits an onset decomposition temperature of 120 °C and a peak decomposition temperature of 147 °C. This is lower than the onset and peak decomposition temperatures of HAN and AMIMAT. The neat HAN shows an onset and peak decomposition temperature of 158 °C and 195 °C and the corresponding values for AMIMAT are 180 °C and 256 °C, repectively. It appears that the formation of a double salt alters the local ionic environment and hydrogen-bonding network, which can increase or decrease the thermal stability depending on composition. In particular, the observed shifts in peak decomposition temperature with changing O/F ratio can be attributed to variations in the extent of double-salt formation and its solubility.

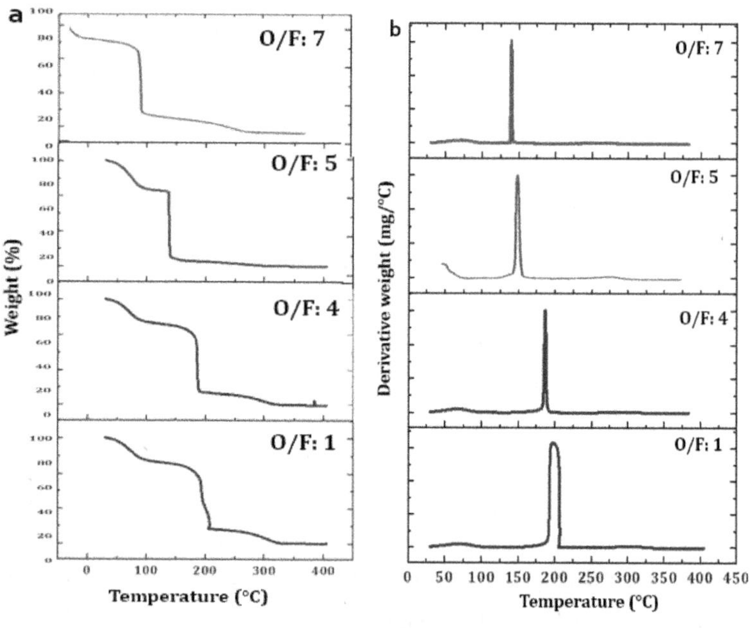

Figure 4.17 (a) TG and (b) DTG of HAN-AMIMAT mixtures at different O/F ratios.

Table 4.5 Tonset and Tpeak decomposition temperatures of HAN-AMIMAT mixtures.

O/F	Tonset (°C)	Tpeak (°C)
1.0	183	196
4.0	180	186
5.0	120	147
7.0	137	139

Differential Scanning Calorimetry (DSC) enables the determination of the endothermic or exothermic nature of thermal events, which are critical indicators of the thermal stability, and handling safety of energetic materials. The DSC analysis of the monopropellant formulations reveals a predominantly exothermic decomposition behaviour across the investigated compositions. The formulation with an O/F ratio of 1 exhibits a peak at a temperature of 207 °C in the DSC. The formulation with an O/F ratio of 5 shows a peak at a significantly lower temperature of 137 °C , suggesting reduced thermal stability and an increased propensity for thermal decomposition at elevated oxidizer content. The observed shift in DSC peak with a change in O/F ratio highlights the strong influence of composition on decomposition kinetics and energy release characteristics, which is crucial for optimizing performance. DSC plot of HAN-AMIMAT at O/F of 1 and 5 is given in **Fig. 4.18.**

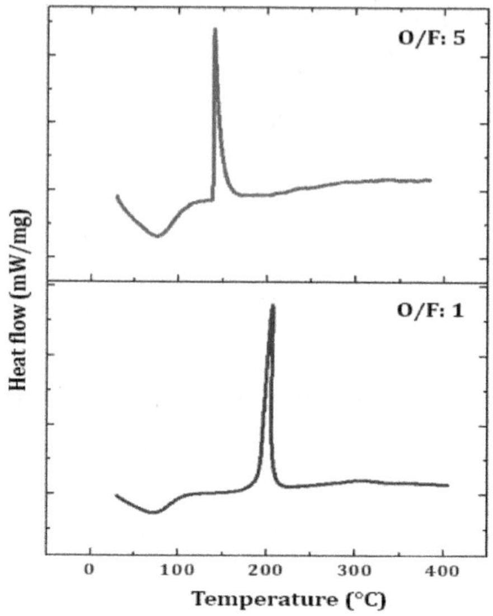

Figure 4.18 DSC plot of HAN-AMIMAT at O/F of 1 and 5.

4.6 Conclusions

The electronic structural parameters of novel EILs based on 1-alkyl-3-methyl imidazolium cation (alkyl=ethyl, butyl, allyl and vinyl) paired with five different anions: nitrate (N), dinitramide (DN), 5-amino-1,2,4-triazolate (Tz), 3,5-diamino-1,2,4-triazolate (DTz), and 5-amino tetrazolate (AT) are theoretically investigated using DFT. 5-amino tetrazolate anion is persuasive in imparting more energetic characteristics when compared to other anions in combination with VMIM cation, but the low HOMO-LUMO energy gap (2.8 eV) makes it more sensitive and less stable, which may pose challenges in handling and safety. In contrast, the allyl-substituted AMIMAT exhibits a high heat of formation (485 kJ/mol) while offering improved stability, making it a more promising fuel candidate.

1-allyl-3-methyl imidazolium-5-amino tetrazolate (AMIMAT), the most endowed candidate with good energetic characteristics is synthesized and characterized for the first time. The onset and peak decomposition temperatures observed for AMIMAT are 180 °C and 256 °C, respectively, and the E_a values for the thermal decomposition is ~121 kJ/mol. Bi-molecular nucleophilic substitution (S_N2) reaction proposed as the mechanism of thermal decomposition based on the decomposition products identified by Py-GC-MS is further substantiated by DFT studies. The HAN-AMIMAT system delivers a maximum vacuum specific impulse of approximately 320 s at an O/F ratio of 5, which is significantly higher than that of 95% HAN. HAN–AMIMAT formulations prepared at different oxidizer-to-fuel (O/F) ratios are evaluated by thermal studies. HAN–AMIMAT formulation at O/F ratio of 5 exhibited single step decomposition with onset and peak decomposition temperatures of 120 °C and 147 °C, respectively. The precipitation observed in the HAN-AMIMAT system at O/F ratios in the range of 1 to 4 due to double-salt formation is substantiated by theoretical studies.

CHAPTER 5

SUBSTITUTED LANTHANUM CHROMITE AS AN EFFICIENT CATALYST FOR HYDROXYL AMMONIUM NITRATE DECOMPOSITION

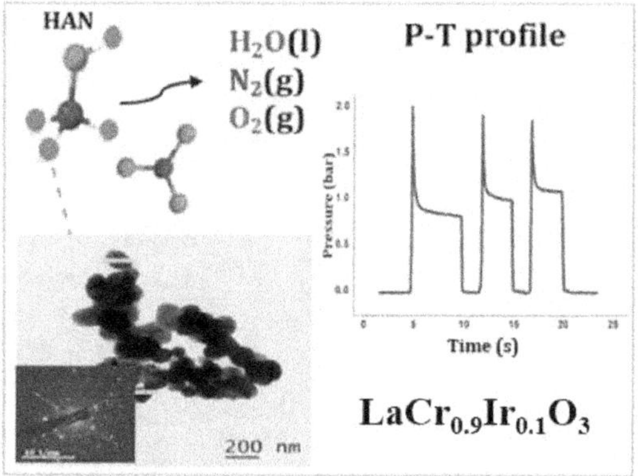

Check for updates

ChemistrySelect

Research Article
doi.org/10.1002/slct.202504030

Chemistry Europe
European Chemical Societies Publishing

www.chemistryselect.org

Cobalt-Doped Lanthanum Chromite as an Efficient Catalyst for the Decomposition of Hydroxyl Ammonium Nitrate (HAN)

Sreelekshmi KR,[a, b] Abhilash KS,[a] Jayalatha T,[a] Vijayalakshmi KP,*[a] and Prabhakaran K*[b]

5.1 Introduction

Monopropellant thrusters power the attitude and reaction control systems of spacecraft and satellites. Hydrazine, a well-established monopropellant, offers excellent performance characteristics, with a specific impulse (Isp) of 234 seconds. However, its high vapor pressure, acute toxicity and carcinogenic nature, creating a significant demand for greener alternatives. Hydroxyl Ammonium Nitrate (HAN) emerges as a promising candidate due to its advantageous properties, including low vapor pressure, low toxicity, high volumetric density (1.4 g/cc), high energy density, and a positive oxygen balance of 33.33% (Amrousse et al., 2015; Hoyani et al., 2017; Katsumi et al., 2012).

The thermal and catalytic decomposition of HAN has been the focus of extensive research over the years. Thermogravimetric analysis (TGA) and differential thermal analysis (DTA) studies indicate that a 95 wt.% aqueous solution of HAN exhibits an onset decomposition temperature of 155 °C, with the decomposition products as H_2O, NO, N_2O, NO_2, and HNO_3 (Amrousse et al., 2012; Katsumi et al., 2015). Catalytic decomposition of monopropellants is the predominant mode of propulsion for satellite thrusters. Katsumi et al. (2015) investigated the catalytic decomposition of a 95 wt.% HAN aqueous solution using differential thermal analysis and thermogravimetry (DTA/TG) with a 32 wt.% iridium catalyst on alumina, observing a significant mass loss and an exothermic peak between 65-75 °C. Additionally, Esparza et al. (2018) conducted experiments using iridium-rhodium foam pellets, achieving a remarkable reduction in decomposition temperature to 95 °C. Mass spectrometry confirmed that the evolved gases are consistent across both thermal and catalytic decompositions, indicating that the catalyst facilitates the initial proton transfer reaction, thereby lowering the decomposition temperature of HAN (Esparza et al., 2018). Amrousse et al. (2012) further explored the catalytic decomposition of 95% HAN using a 10 wt.% iridium catalyst on alumina through thermal analysis and constant volume batch reactor tests, reporting a notable decrease in decomposition temperature to 108 °C, particularly pronounced with concentrated HAN. In another study, an Ir-CuO bimetallic catalyst on a honeycomb structure is proposed as a cost-effective and long-lasting alternative to the Shell catalyst (Amrousse et al., 2017). However, hydrazine decomposition catalysts, such as Shell 405-Ir/Al_2O_3, are unsuitable for HAN due to its significantly higher adiabatic flame temperature (over 2000 °C compared to 900 °C for hydrazine). This elevated flame temperature, combined with the highly corrosive environment, presents substantial challenges in developing effective catalysts for HAN decomposition.

Iridium-based catalyst systems have been widely employed for monopropellant decomposition since the development of the conventional Ir/Al_2O_3 catalyst. Despite their effectiveness, the use of noble metals like iridium presents major limitations due to their high cost (Zhou et al., 2025). Consequently, there are growing research efforts increasingly directed towards developing catalysts with reduced iridium loading or exploring low-cost, non-noble metal substitutes that can deliver comparable catalytic activity and stability. The multivalent oxidation states of cobalt (Co) and cerium (Ce) oxides (Co(II) and Co(III), Ce(III) and Ce(IV)) are utilized in the catalytic decomposition of HAN (Agnihotri and Oommen, 2020). Perovskite-type rare earth mixed metal oxides, with the general formula ABO_3 (where A represents a rare earth ion and B denotes a transition metal cation), have gained significant attention due to their broad applications in catalysis and solid oxide fuel cells (Humayun et al., 2022; Peña and Fierro, 2001; Tejuca et al., 1989). Lanthanum chromite perovskite ($LaCrO_3$) is notable for its high stability, thermal conductivity, mechanical strength, and elevated melting points (~2500 °C), along with relatively high catalytic oxidation activity (Madoui and Omari, 2016). Substituents such as cobalt ($LaCr_{1-x}M_xO_3$, M=Co), enhance the mechanical, chemical, and thermal stability of lanthanum chromite (Adaika and Omari, 2015; Chadli et al., 2016; Madoui and Omari, 2016; Wellia et al., 2023). Moreover, the surface defects and oxygen vacancy accord for the catalytic active sites in perovskite oxides. Although numerous studies have reported on the application of lanthanum-based perovskites as catalysts, their use specifically as monopropellant catalysts remains limited (Hunt et al., 2015).

The perovskites are synthesized through various methods, including solid-state reactions (Coskun et al., 2019), solution combustion methods, sol-gel route

(Silva et al., 2021) and hydrothermal method (Rivas-Vázquez et al., 2006). The choice of preparation route significantly impacts parameters such as crystallinity and surface area which are important for catalytic activity. Among these, the solution combustion method is favoured for synthesising multicomponent oxides due to its simplicity and ability to achieve good homogeneity (Branquinho et al., 2016; Deshpande et al., 2003; Siddique et al., 2022; Varma et al., 2016; Zhu et al., 2013). Different chemicals such as urea, glycine, glucose, etc are used as fuel in combustion synthesis to produce metal oxide particles with varying sizes and morphologies (Akbi et al., 2024). The morphology of synthesized nanoparticles can be fine-tuned using capping agents or surfactants, which selectively adsorb onto the particles, altering their surface free energy. Utilizing natural capping agents derived from plant extracts for nanoparticle synthesis is appealing due to environmental sustainability and cost-effectiveness.

Several studies (Munirathnam et al., 2023; Nasir Ahamed et al., 2024; Shetty et al., 2022; Vasudha et al., 2021) have demonstrated the use of aloe vera gel as a reducing and stabilising agent in the synthesis of various oxide nanoparticles via solution combustion method. The present work focuses on the synthesis of substituted lanthanum chromite using the solution combustion method, employing urea as the fuel and aloe vera gel as a capping agent to regulate particle size and morphology. The application of aloe vera in perovskite synthesis, yielding nanoparticles effective for monopropellant decomposition, is reported here for the first time. The effectiveness of the synthesized materials in catalyzing HAN decomposition is examined through both thermal analysis and batch reactor testing.

5.2 Experimental

5.2.1 Synthesis of catalyst

Analytical reagent grade Lanthanum nitrate ($La(NO_3)_3.6H_2O$), Chromium nitrate ($Cr(NO_3)_3.9H_2O$), Cobalt nitrate ($Co(NO_3)_3.6H_2O$), Hydrogen hexachloroiridate (IV) hydrate ($H_2Cl_6Ir \cdot xH_2O$) and urea were procured from Ottochemie, India. All chemicals were used as such without further purification. The deionized water was used for the preparation of solutions. Aloe vera gel was extracted from the plant leaf after removing the green skin. The gel was ground in a mixer grinder to form a colloidal solution, which was filtered through a Whatman 42 filter paper and used. The solid content was estimated gravimetrically by drying a known weight of the aloe vera gel in an air oven at 90 °C to a constant weight. The solid content obtained was 0.85 wt.%.

$LaCr_{1-x}Co_xO_3$ and $LaCr_{1-x}Ir_xO_3$ perovskites were synthesized by the solution combustion method. Typically, for the synthesis of $LaCr_{1-x}M_xO_3$ (x =0-0.5) for x =0.2 (M=Co/Ir), 1 mmol of $La(NO_3)_3 \cdot 6H_2O$, 0.8 mmol of $Cr(NO_3)_3 \cdot 9H_2O$ and 0.2 mmol of $Co(NO_3)_3 \cdot 6H_2O$ or $H_2Cl_6Ir \cdot xH_2O$ were dissolved in 20 ml water. 12 ml of the aloe vera gel extract was added to the metal nitrate solution and thoroughly mixed by stirring for 2 hours. Subsequently, the urea fuel was dissolved in the solution. The urea to metal nitrate mole ratio was kept at 5. The solution was then heated on a hot plate at 80 °C with constant stirring to obtain a dry mass. The dry mass obtained was calcined in a muffle furnace at 800 °C in air atmosphere for 4 h at a heating rate of 5 °C/min to get the $LaCr_{1-x}M_xO_3$ (M=Co/Ir) powder. A similar procedure was performed without the aloe vera gel for comparison. Cobalt substituted lanthanum chromite synthesized with and without aloe vera gel are abbreviated as LCCO and LCCO-AV, respectively. Similarly, the Ir-substituted lanthanum chromite synthesized with aloe vera gel is denoted as LCIO-AV. A schematic of the solution combustion synthesis is shown in **Fig.5.1**

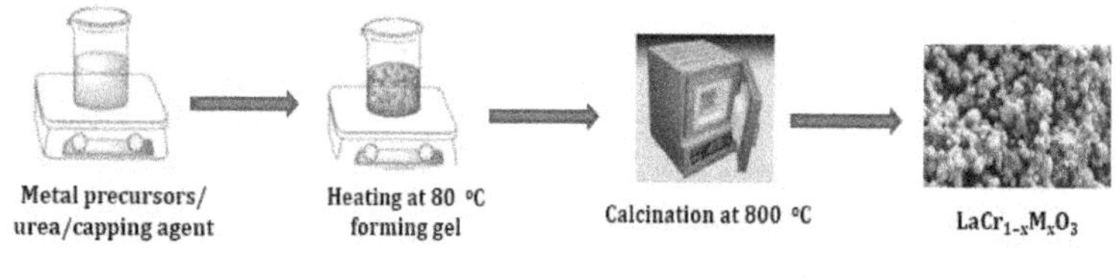

Figure 5.1 Schematic of the synthesis of the catalyst.

5.2.2 Characterization

***FTIR Analysis*:** FTIR analysis was performed using Nicolet iS50 FTIR spectrometer covering a wavenumber range from 4000 to 400 cm^{-1} with KBr pellets. In order to take spectrum, 0.1 to 1.0% of the sample is mixed with fine KBr powder using an agate mortar.

***XRD Analysis*:** X-ray diffraction is used for the determination of the atomic and molecular structure of a crystal, in which the atoms in the crystalline material cause a beam of incident X-rays to diffract into many specific directions. By measuring the angles and intensities of these diffracted beams, one produces a three-dimensional picture of the density of electrons within the crystal. From this electron density, information such as the mean positions of the atoms in the crystal, chemical bonds, and bond disorder, can be obtained. Using this technique, we can get the information regarding the crystalline nature of a material, nature of the phases present, the lattice parameter and grain size. The interaction of X-ray radiation with the crystalline sample is governed by Bragg's law ($n \lambda = 2d\sin \theta$), which depicts a relationship between the diffraction angles (Bragg angle), X-ray wavelength, and interplanar spacing of the crystal planes.

X-ray diffraction (XRD) was performed on a Bruker D-8 Discover X-ray diffractometer, utilizing monochromatic CuKα radiation (1.54 Å) at a scan rate of 2.5°/min. Crystalline phases were identified by comparing the experimental diffraction patterns to Joint Committee on Powder Diffraction Standards (JCPDS). The crystallite size was calculated from the XRD data using Scherrer equation, $D = k \lambda / \beta\cos \theta$, where D is the crystallite size, k is the Scherrer constant, λ is the wavelength of the x-ray, β is the full width half maximum and θ is Bragg's angle.

***XPS Analysis*:** XPS is based on the photoelectric effect that identifies the elemental composition, chemical state and density of the electronic states in a material. XPS analysis involves irradiating a sample with monoenergetic X-rays under high vacuum. When the X-ray interacts with the core level electrons of the surface atoms, the electron absorbs the photo energy. However, if this energy is high enough to overcome the binding energy, the electron from the atom is ejected as a photoelectron. The spectrometer measures the kinetic energy of the emitted electron and binding energy is calculated as:

$$E_b = E_p - (E_k + \Phi),$$

where E_b: Binding Energy of the electron, E_p Energy of the incident X-ray, E_k: Kinetic Energy of the emitted electron and Φ: Work function of the spectrometer.

X-ray Photoelectron Spectroscopy (XPS) spectra were obtained with Thermo Scientific™ ESCALAB™ Xi+ using Al Kα X-rays source, calibrated according to the C1s peak at 284.5 eV.

***ICP-OES analysis*:** The elemental composition was analysed using a Thermo Scientific iCap 7000 series ICP-OES having a segmented array CCD detector. 0.1 g of LCCO-AV/ LCIO-AV sample taken in a Pt crucible was fused with 8 g of sodium carbonate by keeping in a muffle furnace at 900°C for 2 hours. Then the mixture was dissolved in HCl and made up to 250ml. A dilute aqueous solution of 1-20 ppm made by diluting the stock solution was fed into ICP-OES. The concentration of La, Cr and Ir in the samples was analysed and quantified from the calibration curve made from the standard solutions at 1ppm, 10 ppm and 20 ppm of the standard material under investigation.

***FESEM Analysis*:** A scanning electron microscope (SEM) is a type of electron microscope that produces images of a sample by scanning it with a focused beam of electrons. The electrons interact with atoms in the sample, producing various signals that can be detected and that contain information about the sample topography and composition. The types of signals produced by a SEM include secondary electrons, back-scattered electrons, characteristic X-rays and transmitted electrons. In the most common or standard detection mode, secondary electron imaging, the SEM can produce very high-resolution images of a sample surface, revealing details less than 1 nm in size. A wide range of magnifications is possible, from about 10 times to more than 500,000 times. Field emission (FE) is emission of electrons induced by an electrostatic field. The most common context is field emission from a solid surface into vacuum. FESEM images and EDS spectra of gold-coated samples were recorded using Zeiss Gemini 500 FESEM at an accelerating voltage of 20 kV. Energy dispersive X-ray spectroscopy (EDS or EDXA) make use of the characteristics X-rays emitted by a solid sample bombarded with a beam of electrons to identify the element and measure its abundance.

***TEM Analysis*:** Transmission electron microscopy (TEM) is a microscopy technique in which a beam of electrons is transmitted through an ultra-thin specimen, interacting with the specimen as it passes through it. An image is formed from the interaction of the electrons transmitted through the specimen; the image is magnified and focused onto an imaging device, such as a fluorescent screen, on a layer of photographic film, or to be detected by a sensor such as a charge-coupled device. Jeol/JEM 2100 High-Resolution Transmission Electron Microscope with an accelerating voltage of 200 kV was used for the analysis. For HRTEM measurements, the sample was dispersed in isopropyl alcohol/ethanol, drop-casted on a carbon-coated copper grid and dried under vacuum at room temperature before observation.

Surface area analysis: Brunauer Emmett Teller (BET) method was used to determine surface area of the samples. The surface area of the catalysts was determined using a TriStar II surface area analyser, applying the BET equation. The samples were degassed at 300 °C in flowing nitrogen for 6 hrs before the analysis.

Evaluation of catalytic activity by thermal analysis:

Differential thermal analysis-thermogravimetric analysis (TA instrument-Q50) and differential scanning calorimeter (DSC, TA instrument Q20) were used to study the activity of the synthesized catalysts. In thermal analysis experiments for the evaluation of activity of the synthesised catalyst towards HAN decomposition, a known mass of catalyst material was placed in the sample pan (alumina pan), followed by the addition of 5-6 mg of HAN and subjected to thermal studies in a flowing 99.99% pure nitrogen environment at a heating rate of 5 °C/min. The initial decomposition temperature (Tonset) and temperature of maximum thermal decomposition (Tpeak) were obtained from the thermogram. A decrease in Tonset and Tpeak in presence of the catalyst was measured as its catalytic activity.

To investigate the decomposition behaviour, a kinetic analysis of HAN decomposition was performed both with and without the catalyst using Differential Scanning Calorimetry (DSC) under the same experimental conditions but at varying heating rates. The activation energy of decomposition was calculated using the

Kissinger method (Kissinger, 1957) and the activation energy E_a is attained from equation:

$$\ln\left(\frac{\beta}{T^2}\right) = \ln\left(\frac{AR}{T}\right) - \frac{Ea}{RT} \qquad (5.1)$$

β is the heating rate, T, the temperature in Kelvin scale at a fixed conversion value of α. A is the pre-exponential factor, E_a is the activation energy at a given conversion and R is the gas constant. For the Kissinger method, $\ln\left(\frac{\beta}{T^2}\right)$ was plotted against $1/T$ giving rise to a straight line whose slope yields activation energy.

TG-MS analysis: The combination of a TG with an MS allows detection of very low levels of volatile products in real time. By heating a sample on the TG instrument, the sample will release volatile products and are transferred to the MS where the components are identified. Perkin Elmer Pyris 1 Thermogravimetric analyser Coupled with Perkin Elmer Clarus SQ8T mass spectrometer was used in tandem for the TG-MS studies for the identification of evolved species from 10 to 500 amu mass range in He atmosphere at a heating rate of 5 °C/min. In the TG-MS mode, TL 200 flexible heated transfer line (1.6 m X 0.1 mm i.d. silica transfer line) was used to connect TG to MS. The transfer line was connected to the TG through the sheath gas inlet. A few centimetres of the deactivated fused silica transfer line protrudes into the sample weighing area of the TG. The other end of the transfer line was connected to the mass spectrometer.

Batch reactor test: The catalytic activity was monitored using a constant volume batch reactor in pulse mode (Agnihotri and Oommen, 2020; Eloirdi et al., 2003). The batch reactor, designed and built in-house was used for HAN decomposition studies. The batch reactor comprises of a small stainless steel-304 chamber in which the pressure and temperature rise are monitored following monopropellant injection. About 0.3 g of catalyst was loaded into the reactor chamber, which was maintained isothermally at 175 °C using a PID-controlled heater mounted beneath the bottom plate. In pulse mode operation, approximately 200 μl of HAN was injected into the preheated catalyst bed and the resulting pressure rise was recorded using pressure transducers to evaluate the decomposition behaviour of the monopropellant.

5.3 Results and Discussion

5.3.1 Co-substituted lanthanum chromite

Substituted lanthanum chromite (ABO_3) perovskites are prepared by the solution combustion method employing urea as fuel. The nature and extent of substitution of the metal in the A and B sites modify the properties. The partial substitution at the B site is reported to improve the perovskite stability and redox activity (Madoui and Omari, 2016). $LaCr_{1-x}Co_xO_3$ with x=0 to 0.5 is prepared and their XRD spectrum is presented in **Fig. 5.2(a)**.

The diffraction peaks observed at 2θ values of 22.95, 32.6, 40.3, 46.8, 52.8, 58.3, 68.4, 73.1 and 77.9 corresponded to the (110), (020), (022), (004), (130), (132), (040), (006), (240), crystallographic planes characteristic of the orthorhombic type structure (Madoui and Omari, 2016). On the other hand, the main diffraction peaks corresponding to (020) plane get split into two in the composition with $x > 0.3$ as shown in **Fig. 5.2(b)**, indicating the rhombohedral crystal geometry. However, in all studied compositions, the perovskite structure is retained without any spinel phase. An increase in the B site doping by Co ion from $x > 0.2$ results in orthorhombic to rhombohedra phase change and a decrease in unit cell parameters. This is attributed to the distortion of coordination in the octahedral arrangement of chromium with an ionic radius of 0.61 Å being substituted with cobalt having a smaller radius of 0.545 Å (Corrêa et al., 2008).

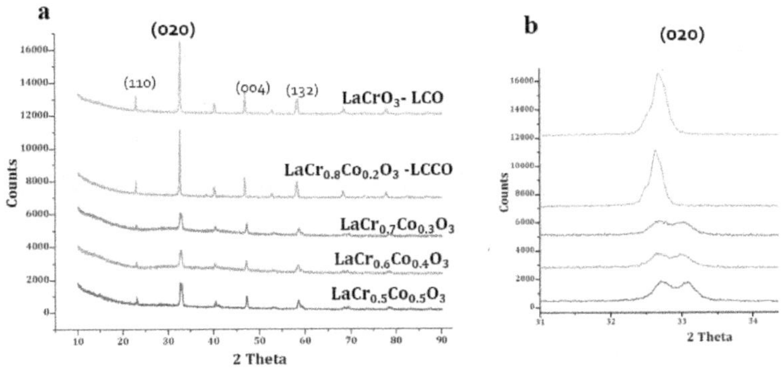

Figure 5.2 (a) XRD pattern and (b) the diffraction peak corresponding to (020) plane of $LaCr_{1-x}Co_xO_3$ with x = 0–0.5.

Even though the introduction of dopants helps to tailor the desired properties of the perovskites, the volume shrinkage and orthorhombic to rhombohedral phase transition associated with increased loading of Co can generate internal stress and can cause mechanical failure due to the wide range of temperature changes experienced during the catalytic decomposition of HAN. Hence $LaCr_{0.8}Co_{0.2}O_3$ (LCCO) having orthorhombic phase is selected as the catalyst for HAN decomposition based on crystallographic studies. It has also been reported that increasing Co doping level generates non-uniform particle size due to particle agglomeration (Coskun et al., 2019).

5.3.1.1 Green capping agent for synthesising $LaCr_{0.8}Co_{0.2}O_3$

The aloe vera gel-assisted synthesis is aimed at producing Co substituted lanthanum chromite crystals with minimal aggregation and controlled morphology (Begum et al., 2025). Aloe vera gel, is reported as an effective natural capping and reducing agent for many nanoparticle synthesis (Majani et al., 2025; Munirathnam et al., 2023). Aloe vera contains phytochemicals such as phenolic acids, polysaccharides and flavonoids. The phenolic acid contains multiple hydroxyl, carbonyl, and methoxy groups enable the reduction of metal ions. The phytochemical compounds, such as flavonoids and polysaccharides, present in the gel facilitate nanoparticle formation by adsorbing onto the particles, thereby restricting their growth and maintaining their spherical shape. Aloe vera gel extracts contain aloin, which causes steric repulsion, preventing nanoparticles from aggregating by decreasing the overall surface energy. Moreover, aloe vera acts as a soft template to fine-tune the nanoparticle shape and distribution by preventing particle agglomeration, controlling crystal growth and maintaining uniform particle size distribution (Khaldoune et al., 2024; Ragupathi et al., 2014). A general schematic showing capping of nanoparticles with phytochemicals from aloe vera gel is shown in **Fig. 5.3**.

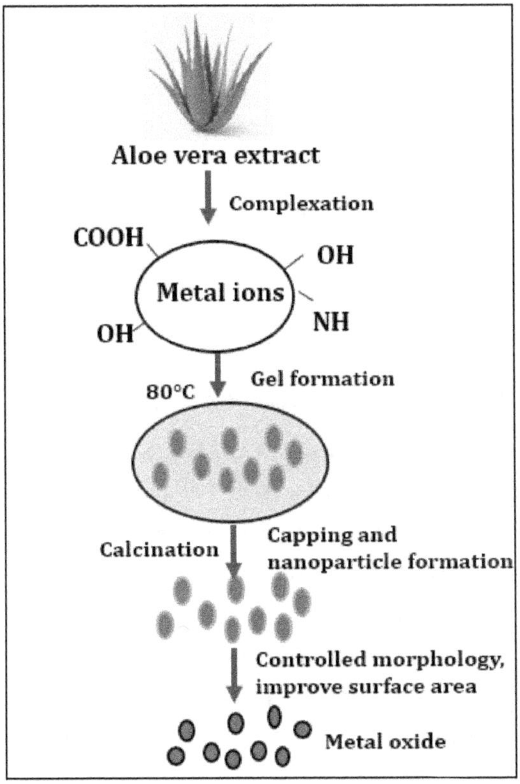

Figure 5.3 Schematic showing the capping of nanoparticles by phytochemicals from aloe vera gel.

XRD patterns of $LaCrO_3$ (LCO), $LaCr_{0.8}Co_{0.2}O_3$ synthesized without (LCCO) and with aloe vera gel (LCCO-AV) are presented in **Fig. 5.4**. The comparison of these spectra with JCPDS charts indicates that all samples prepared are perovskites type mixed oxides and exhibit XRD pattern similar to that of $LaCrO_3$. The crystallite size of the perovskites was calculated from XRD data by Scherrer equation. The crystallite size obtained for LCO, LCCO and LCCO-AV are 44 nm, 37.2 nm and 36.5 nm, respectively. The decrease in crystallite size can be attributed to the substitution of Cr ions with Co ions of smaller radius.

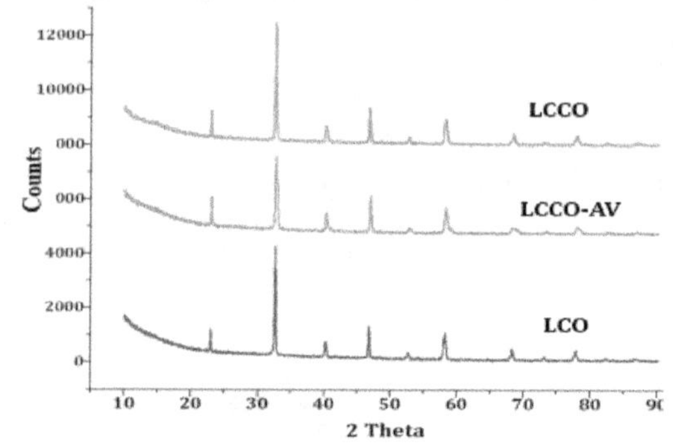

Figure 5.4. XRD pattern of LCO, LCCO and LCCO-AV.

FTIR spectra of LCO and LCCO-AV are illustrated in **Fig. 5.5.** The vibrational frequencies observed in the range 700-400 cm^{-1} correspond to the MO_6 octahedra in the perovskite phase (Madoui and Omari, 2016; Zarrin et al., 2019). The high frequency absorption bands at around 611 cm^{-1} arise from La-O, Cr-O or Co-O stretching vibrations and the low-frequency band at 418 cm^{-1} are attributed to M-O-M bending mode of vibrations. Additionally, absorption bands at 3433 cm^{-1} and 3439 cm^{-1} are ascribed to the stretching vibrations of O-H bonds in adsorbed water molecules.

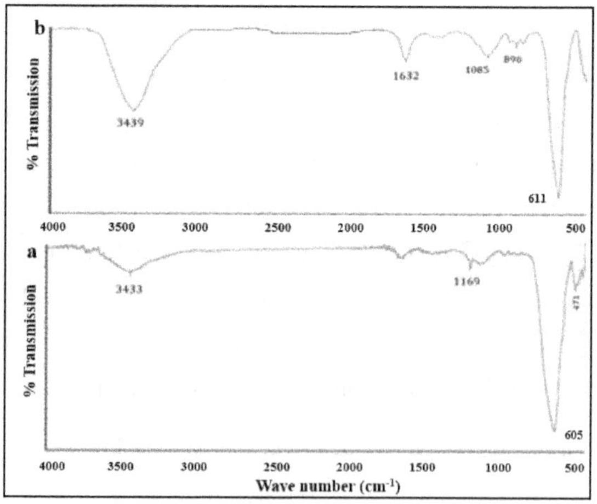

Figure 5.5 FTIR spectrum of (a) LCCO-AV and (b) LCO.

The preparation route selected for the synthesis of perovskites significantly influences their morphological and structural features. In the solution combustion method, a precursor solution containing metal nitrates (as oxidizers) and urea (as fuel) undergoes thermal treatment, leading to dehydration and decomposition through exothermic reactions that produce large volumes of gaseous by-products, resulting in solid metal oxide products. The surface morphology of these solid products, evaluated by FESEM, is depicted in **Fig. 5.6**.

Figure 5.6 FESEM of (a) LCO, (b) LCCO and (c) LCCO-AV.

Both LCO and LCCO exhibit highly aggregated particles with irregular shapes, with LCO displaying uneven particle sizes ranging from several hundred nanometres to one micron. Cobalt substitution contributes to a reduction in particle size, with LCCO particles measuring between 120 and 600 nm. In contrast, LCCO-AV features distinct spherical particles with a narrower size distribution, ranging from 70 to 350 nm. This reduction in particle size for LCCO-AV is attributed to the capping effect of aloe vera gel. The phytochemical compounds, such as flavonoids and polysaccharides, present in the gel facilitate nanoparticle formation by adsorbing onto the particle surface, thereby restricting their growth and maintaining their spherical shape. Consequently, the aloe vera gel-assisted synthesis of LCCO-AV is aimed at producing crystals with minimal aggregation and controlled morphology (Begum et al., 2025). The observed differences in the particle size by XRD and SEM analysis are due to the clustering of nanoparticles due to molecular structural disorder and lattice strain (Ragupathi et al., 2014).

The elemental mapping of LCCO-AV shown in **Fig. 5.7(a)** indicates the homogenous distribution of La, Cr, Co and O in the sample. The EDX spectrum of

LCCO-AV is shown in **Fig. 5.7(b)**. The EDX spectrum confirms the La, Cr, Co and O in LCCO-AV and did not show any impurity phase arising from aloe vera gel. The percentage of La, Cr, Co and O obtained from the EDX analysis is 59.78, 19.89, 4.19 and 16.14 wt.% respectively. On the other hand, the percentage of La, Cr, Co and O obtained from ICP-OES analysis are 56.9, 16.7, 4.7 and 21.7 wt.% respectively. The elemental composition obtained from EDX and ICP-OES analysis are close to the theoretical composition of LCCO-AV viz: 57.8 wt.% La, 17.3 wt.% Cr, 4.9 wt.% Co and 19.98wt.% O.

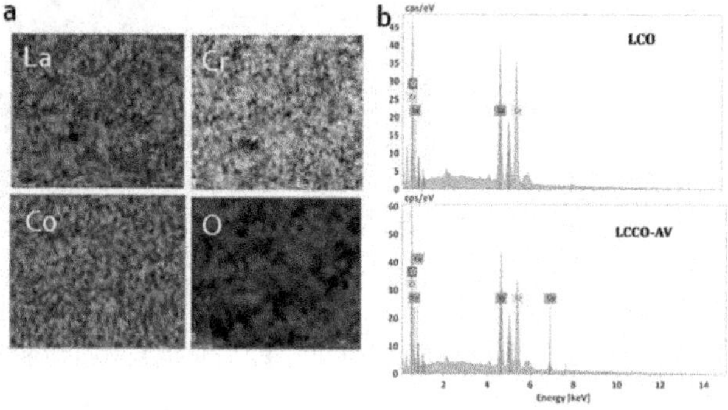

Figure 5.7 (a) Elemental mapping and (b) EDX spectra of LCCO-AV.

The survey and deconvoluted XPS spectra of LCO and LCCO-AV are illustrated in **Fig. 5.8(a)**. Analysing the oxidation states of oxygen and the B-site metal atom is crucial, as these factors significantly influence catalytic activity. The binding energy of La 3d for LCO and LCCO is approximately 834 eV and 852 eV, respectively, confirming the trivalent state of La ions, which remains unchanged with B-site substitution (**Fig. 5.8(b)**). The peak at a binding energy of 579.6 eV shows the presence of chromium in +5/+6 oxidation state which appear typically at the same binding energy values (Rida et al., 2007). The peak at 576 eV is attributed to the presence of Cr in the +3 state. The Cr 2p spectrum of LCCO and LCCO-AV are presented in **Fig.5.8(c)**.

In LCCO-AV, the concentration of chromium in +5/+6 oxidation state is more, since the intensity of peak at 579.6 eV is higher in this case. The presence of Cr in higher oxidation states in LCCO-AV is for maintaining electrical neutrality, and this is further confirmed by XRD results. The XRD spectrum does not show any evidence for the presence of phases such as $LaCrO_4$ or La_2CrO_6 with Cr in the +5/+6 oxidation states. This suggests that Cr in the +5/+6 oxidation states is for maintaining the electroneutrality as a consequence of Co substitution (Erdenee et al., 2017; Peron and Glisenti, 2019). O1s spectrum of LCO and LCCO-AV investigated is represented in **Fig. 5.8(d)**. The O1s peak in the range 528-529 eV is ascribed to lattice O^{2-} and the peak between 530-532 eV is due to adsorbed O^- / OH^- groups. The intensity of lattice oxygen and adsorbed oxygen increases in LCCO-AV due to Co doping. **Fig. 5.8(e)** shows the Co 2p spectrum of LCCO-AV. Cobalt ion shows mixed oxidation states of +2 and +3 with peaks at 780 eV (2p3/2) and 795.6 eV (2p1/2), respectively(Coskun et al., 2019).

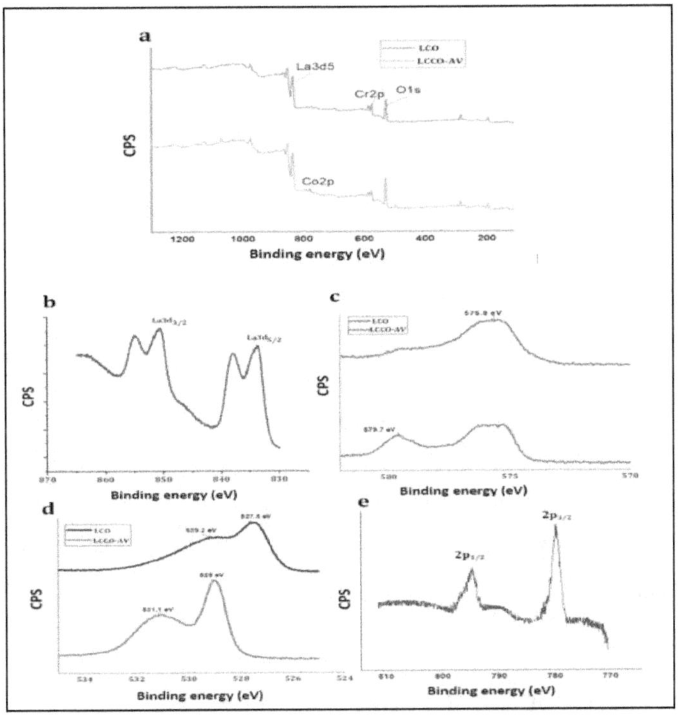

Figure 5.8 XPS plot (a)Survey spectra of LCO and LCCO-AV (b) La3d (c) Cr 2p (d) O1s and (e) Co 2p for LCCO-AV.

The nitrogen adsorption isotherm of LCO and LCCO-AV perovskites belongs to Type III in the International Union of Pure and Applied Chemistry (IUPAC) classification (Yoo et al., 2024), which means the adsorption took place on the surface of a nonporous or macroporous solid with not well-developed micropores. Nitrogen adsorption isotherms of LCO, LCCO and LCCO-AV are shown in **Fig. 5.9**. Cobalt-substituted lanthanum chromite shows a lower surface area of 2.7 m^2/g compared to 3.3 m^2/g for the un-substituted sample. However, LCCO-AV shows a surface area of 4 m^2/g. This slight improvement in surface area can be due to the decrease in particle size arising from the capping effect of the aloe vera gel. The surface area of LCO, LCCO and LCCO-AV are presented in **Table 5.1.**

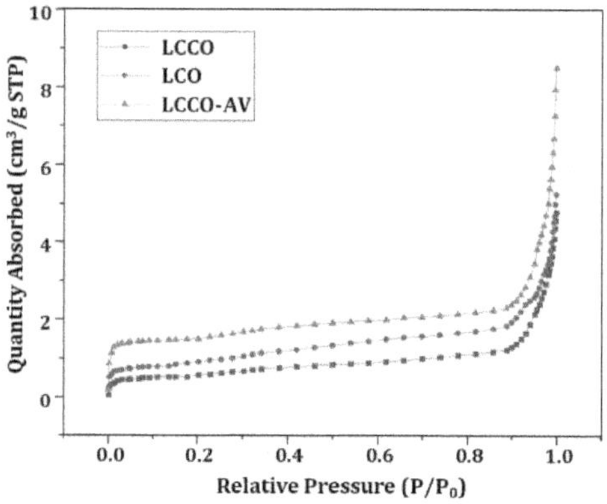

Figure 5.9 Nitrogen adsorption isotherms of LCO, LCCO and LCCO-AV.

Table 5.1 Surface area from BET analysis.

Catalyst	LCO	LCCO	LCCO-AV
Surface area (m^2/g)	3.3	2.7	4.0

5.3.1.2 Evaluation of catalytic activity

The catalytic activity of LCO, LCCO and LCCO-AV towards HAN decomposition is studied using TGA (Shamjitha and Vargeese, 2023). The thermal analysis provides information regarding evaporation of water from the HAN solution indicated by a weight loss in TGA at temperature below 100 °C. The onset and peak decomposition temperatures are obtained from the DTG graphs. The efficiency of the catalyst is indicated by the sharpness of the exothermic peak corresponding to the HAN decomposition. The mass spectrometer coupled with the TGA identify the decomposition products. The catalytic activity is primarily assessed by the lowering of the onset and peak decomposition temperatures. The onset and peak decomposition temperatures of HAN with the perovskites as a catalyst is notably lower than the value obtained for the pristine HAN as shown in **Fig. 5.10.**

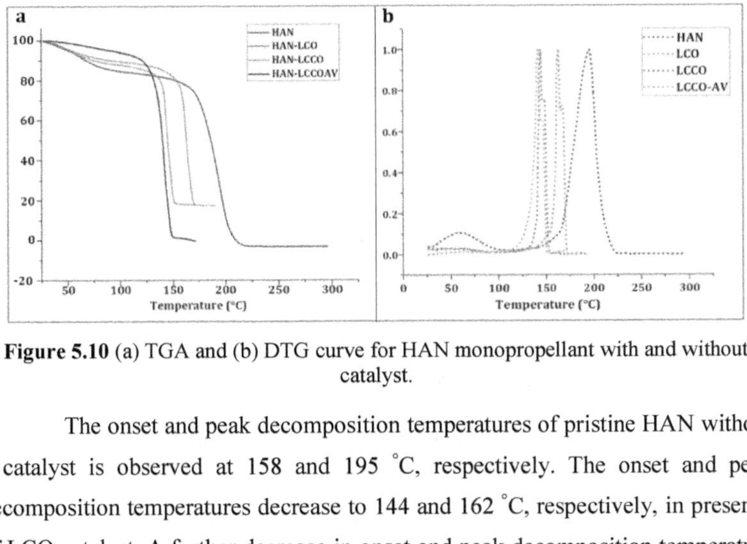

Figure 5.10 (a) TGA and (b) DTG curve for HAN monopropellant with and without catalyst.

The onset and peak decomposition temperatures of pristine HAN without a catalyst is observed at 158 and 195 °C, respectively. The onset and peak decomposition temperatures decrease to 144 and 162 °C, respectively, in presence of LCO catalyst. A further decrease in onset and peak decomposition temperature to 131 and 143.7 °C, respectively, is achieved with LCCO, indicating the enhanced catalytic activity by substitution of Cr with Co. The catalytic activity of perovskite oxides depends on the B site ion since the A site is catalytically inactive and contributes only to stability (Yang et al., 2022). Indeed, the catalytic activity of the perovskites depends primarily on the redox properties of the B-site metal cation. TGA data reveal a significant enhancement in catalytic activity for cobalt substituted perovskites probably due to the presence of oxygen vacancies in the lattice of LCCO (Höfer and Kock, 1993) and redox properties of Cr and Co ions

(Zhu et al., 2014). On the other hand, HAN exhibit a still lower onset and peak decomposition temperatures in the presence of LCCO-AV catalyst. The onset and peak decomposition temperatures observed are 119 and 141 °C, respectively. The improved catalytic performance of LCCO-AV can be attributed to the enhancement in surface area due to the aloe vera gel capping agent (Erdenee et al., 2017). The steeper slopes observed in the TGA curve of HAN with LCCO-AV correspond to an improved rate of reaction.

While heating, the decomposition of HAN will initiate only after the evaporation of the water as evidenced by the shallow peak in the DTG curve below 100°C. Agnihotri et al studied the decomposition of HAN using Ir/CeCo 26 and Ir/γ-Al_2O_3 catalyst and compared the results (Agnihotri and Oommen, 2020). The peak decomposition temperature obtained for Ir/γ-Al_2O_3 and Ir/CeCo 26 catalyst systems are 119.7±2 and 126.1 ±1.3 °C, respectively. Even though the peak decomposition temperature observed in the present study is higher than that reported earlier using Pt and Ir based catalyst system, the low cost and high temperature stability of LCCO-AV is promising for practical application in HAN based propulsion system. The performance of the perovskite catalysts synthesized in this study in terms of lowering of onset and peak decomposition temperatures of HAN is provided in **Table 5.2** in comparison with that of Ir/γ-Al_2O_3 and Ir/CeCo 26 catalyst reported in the literature.

Table 5.2 Comparison of the onset and peak decomposition temperatures of HAN decomposition using different catalysts.

Parameter	HAN alone	[c]LCO	[c]LCCO	[c]LCCO-AV	Ir/γ-Al_2O_3	[a]Ir/CeCo 26
Tonset (°C)	158	144	131	119	[b]42	-
Tpeak (°C)	195 [a]199.5± 2.2	162.7	143.7	141	[a]119.7±2, [b]111	[a]126.1 ±1.3

[a] (Agnihotri and Oommen, 2020), [b] (Amrousse et al., 2012)
[c] Catalyst to HAN w/w ratio in every sample is kept 1:3 by weight.

To further support on the catalytic effect of LCCO-AV on HAN decomposition, DSC analysis is conducted. DSC plot of HAN and HAN-LCCO AV is shown in **Fig. 5.11(a)**. The sharp exothermic decomposition observed in the DSC plot in the presence of LCCO-AV shows the enhanced performance of the system. The activation energy of decomposition is determined from the slope of ($\ln(\beta/T^2)$ vs 1/T plots as shown in **Fig. 5.11(b)**. The activation energy of HAN decomposition in the absence of a catalyst agrees with the literature data of 64.28 kJ/mol (Esparza et al., 2018). On the other hand, in the presence of LCCO-AV catalyst the activation energy of HAN decomposition decreases to 56.41 kJ/mol.

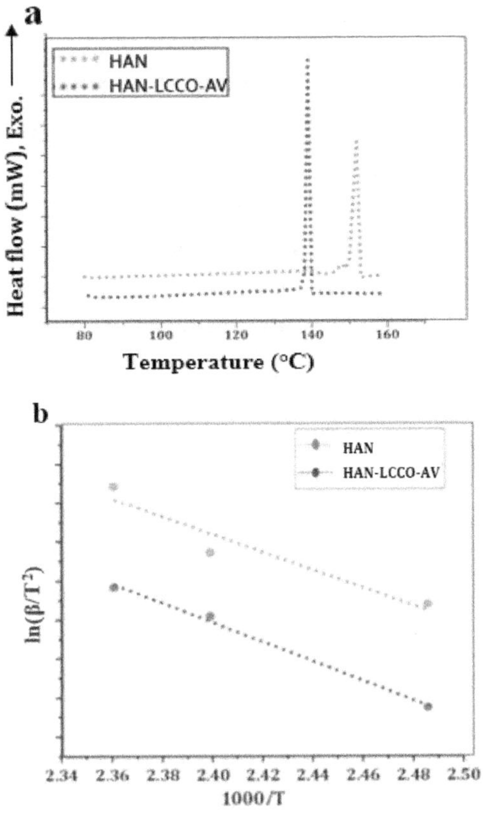

Figure 5.11 (a) DSC plot and (b) Kissinger Plot (($\ln(\beta/T^2)$ vs 1000/T) for HAN and HAN-LCCO-AV.

The performance of the developed catalyst is further evaluated using a batch reactor, where pressure fluctuations during the decomposition of the propellant are monitored in pulse mode operation. The pressure increase obtained with HAN solution indicates a more efficient catalytic reaction and a higher decomposition rate. The pressure-time curve obtained for HAN decomposition with LCCO-AV catalyst evaluated for pulse-mode injections is shown in **Fig. 5.12(a)**. A rapid increase in pressure to a high value is observed in presence of the catalyst. The sharp pressure increase is due to the rapid formation of gaseous products attributed to the following reactions (Amrousse et al., 2012):

$$NH_3OH\ NO_3 \rightarrow 2H_2O(l) + N_2(g) + O_2(g)$$
$$NH_3OH\ NO_3 \rightarrow 2H_2O(l) + 0.5\ N_2(g) + NO_2(g)$$
$$NH_3OH\ NO_3 \rightarrow 2H_2O\ (l) + 2NO(g)$$
$$NH_3OH\ NO_3 \rightarrow 2H_2O(l) + N_2O(g) + 0.5O_2(g)$$

The ignition delay associated with HAN decomposition using the LCCO-AV catalyst is observed to be 0.4 s, which is lower than that observed for the decomposition of HAN without catalyst (14.5 s). **Fig. 5.12(b)** shows the comparison of the pressure profile from batch reactor test without catalyst and with catalyst for the decomposition of HAN.

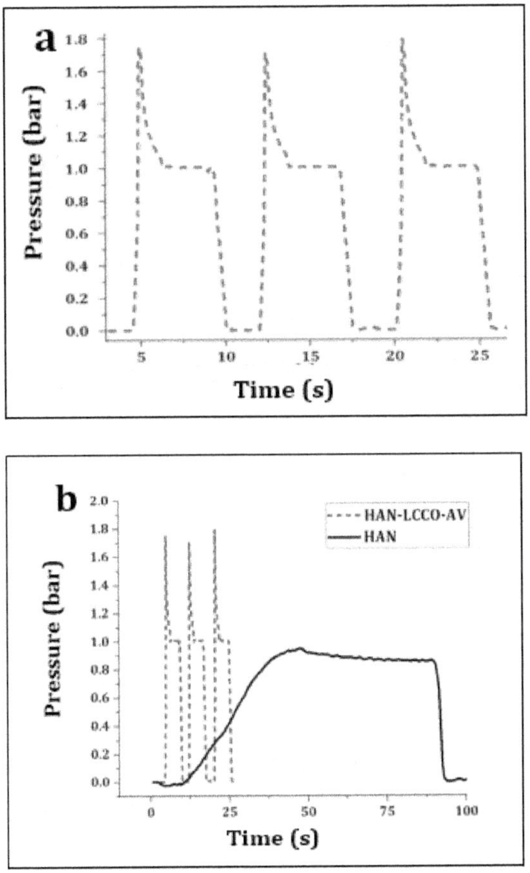

Figure 5.12 Pressure-time profile of (a) HAN with LCCO-AV and (b) Comparison of pressure profile without catalyst and with LCCO-AV.

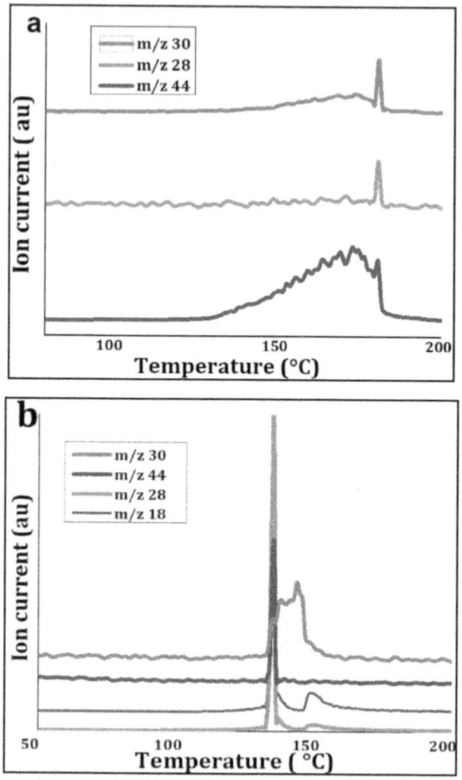

Figure 5.13 Evolved gases during the decomposition of (a) HAN and (b) HAN-LCCO-AV.

The analysis of evolved gases using a mass spectrometer coupled with the TGA reveals peaks corresponding to the molecular mass (m/z) of 30, 44, and 28, indicating the formation of NO, N_2O and N_2 molecules during the thermal decomposition of HAN in the absence of a catalyst. Thermal decomposition of HAN commences only after complete evaporation of water from HAN solution (Chai et al., 2022) and due to the gradual release of water, it is not detected in the mass spectrometer. The evolved gas analysis on HAN decomposition with and without catalyst is shown in **Fig. 5.12(a)** and **Fig.5.12(b)**. Conversely, when the LCCO-AV catalyst is present, the decomposition products include water along with NO, N_2O, N_2. Nitric acid and its products on reaction with hydroxylamine have major role in the formation of intermediate products and the overall decomposition mechanism of HAN. The evolution of N_2O and N_2 occurs through the reactions described in Eqs. (5.4) and (5.5), which involve the interaction of hydroxylamine with the HNO and HNO_2, respectively (Oxley and Brower, 1988).

$$NH_3OH^+ NO_3^- \leftrightarrow NH_2OH+HNO_3 \quad (5.2)$$

$$NH_2OH+HNO_3 \rightarrow HNO+HNO_2+H_2O \quad (5.3)$$

$$NH_2OH+HNO_2 \rightarrow N_2O+2H_2O \quad (5.4)$$

$$NH_2OH+HNO \rightarrow N2+2H_2O \quad (5.5)$$

The formation of NO can be explained by Eq. (5.6) as proposed by Lee and Litzinger (2003) :

$$3HNO_2 \rightarrow 2NO + HNO_3 + H_2O \quad (5.6)$$

The absence of HNO_3, HNO_2 and HNO in the evolved gas indicates that they are produced as intermediates and completely converted into N_2, NO, N_2O and H_2O. Moreover, the decomposition of HAN in the presence of LCCO-AV does not produce acidic NO_2 unlike the proposal by Oxley and Brower (1988). That is, the decomposition in presence of LCCO-AV catalyst yields comparatively harmless products. Moreover, the higher concentration of N_2 suggests a higher performance characteristic due to the evolution of low-molecular-weight species during catalytic decomposition. The presence of cobalt ion in variable oxidation states will also have a distinct influence on catalytic reactivity (Agnihotri and Oommen, 2018). Based on the reported mechanism of reaction of HAN with Plutonium and Iron ions (Barney and Duval, 2011) and Cerium ion (Agnihotri and Oommen, 2018), the observed catalytic activity of LCCO-AV can be attributed to the following proposed redox mechanism given in Eqs. (5.7) and (5.8).

$$Co^{3+} + NH_3OH^+ \rightarrow Co^{2+} + \tfrac{1}{2} N_2 + 2H^+ + H_2O \quad (5.7)$$

$$Co^{2+} + 3/2H^+ + \tfrac{1}{2} NO_3^- \rightarrow Co^{3+} + \tfrac{1}{2} HNO_2 + \tfrac{1}{2} H_2O \quad (5.8)$$

The study suggests that the LCCO-AV has the potential to serve as a cost-effective catalyst for the decomposition of eco-friendly HAN monopropellants.

5.3.2 Ir-substituted lanthanum chromite

Iridium is a well-known and highly active catalyst for the decomposition of hydroxylammonium nitrate (HAN). To investigate the influence of Ir incorporation on the structural characteristics of lanthanum chromite perovskite, iridium-substituted lanthanum chromite with the composition $LaCr_{1-x}Ir_xO_3$ (x= 0.1 to 0. 2) is synthesised by solution combustion method using urea as a fuel and aloe vera gel capping agent. The XRD pattern of the sample with x = 0.2 exhibits a main diffraction peak at $2\theta \approx 32.7°$, characteristic of the orthorhombic perovskite phase of $LaCrO_3$, confirming the formation of the perovskite structure. However, in addition to the perovskite reflections, diffraction peaks corresponding to segregated IrO_2 phase is also observed for compositions with x > 0.1. Even though the ionic radii of Cr^{+3} (0.61 Å) is very close to that of $Ir^{+4,}$ (0.625 Å) solid solution is not obtained beyond x= 0.1. Hence, an optimal iridium-substituted lanthanum chromite, $LaCr_{0.90}Ir_{0.1}O_3$ (LCIO-AV) with x = 0.1, is synthesised and used for further characterisation and catalytic evaluation. **Fig. 5.14** shows the XRD spectrum of $LaCr_{1-x}Ir_xO_3$. The crystallite size of $LaCr_{0.90}Ir_{0.1}O_3$ calculated by the Scherrer equation is 42.5 nm.

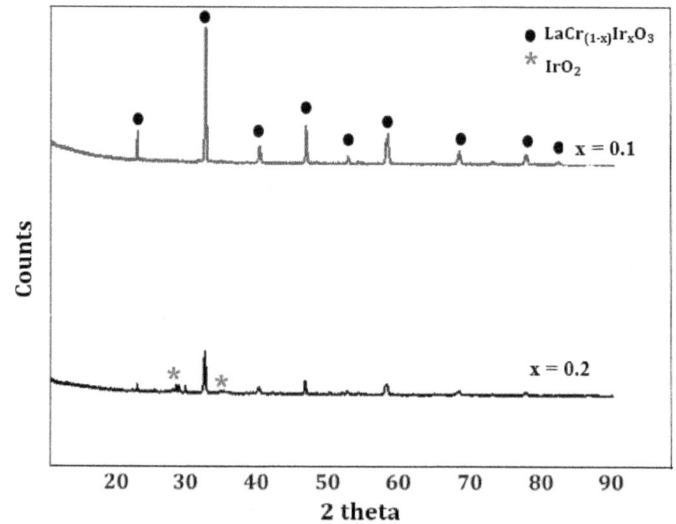

Figure 5.14 XRD pattern of $LaCr_{1-x}Ir_xO_3$ with x=0.1 & 0.2.

The XPS spectra of LCIO-AV is illustrated in **Fig. 5.15**. The binding energy of La 3d for LCIO-AV confirms the trivalent state of La ions (La 3d5/2 [834.1 and 837.8 eV], La 3d3/2 [855.1 and 850.6 eV]), which remains unchanged with Ir substitution. The peak at a binding energy of 579.5 eV shows the presence of chromium in +5/+6 oxidation state which appear typically at the same binding energy values (Rida et al., 2007). The peaks at 575.8 eV (Cr 2p3/2) and 585.7 eV (Cr 2p1/2) are attributed to the Cr in the +3 oxidation state. O1s spectrum of LCIO-AV at 529.1 eV is ascribed to oxygen in perovskite framework and the peak at 531 eV is due to adsorbed O^- / OH^- groups. Ir 4f for LCIO-AV confirms the existence of Ir in the +4 state with binding energies at 64.9 and 62.1 eV.

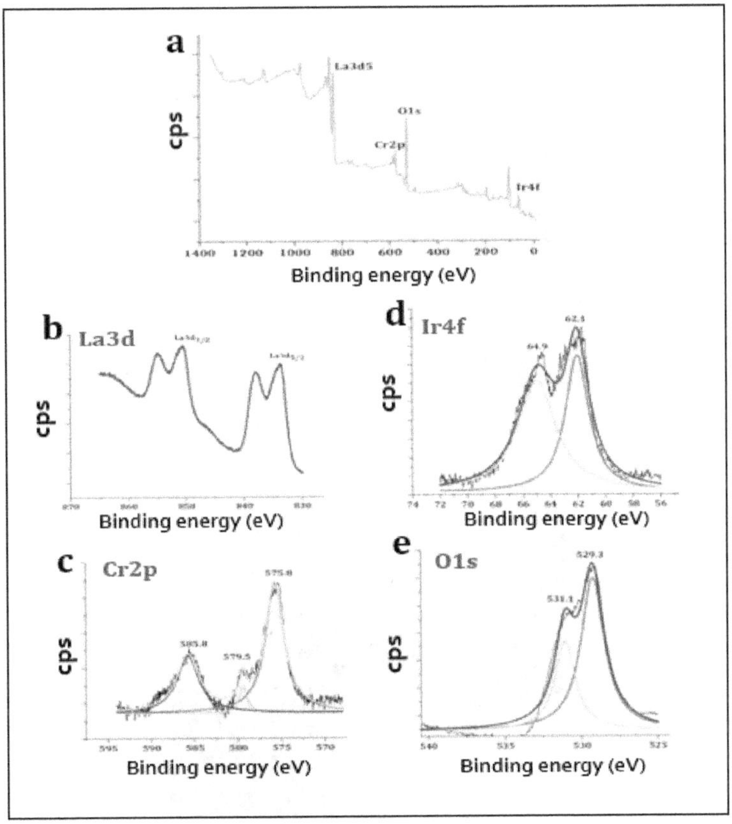

Figure 5.15 XPS plot (a)Survey spectra (b) La3d (c) Cr 2p (d) Ir 4f and (e) O1s for LCIO-AV.

The surface morphology of LCIO-AV presented in **Fig. 5.16(a)** shows highly aggregated particles with irregular shapes, with particle sizes ranging from nanometres to one micron. The EDX spectrum of LCCO-AV is shown in **Fig. 5.16(b)**. The EDX spectrum confirms the La, Cr, Ir and O in LCCO-AV and did not show any impurity phase arising from aloe vera gel. The theoretical elemental composition of LCIO-AV is 54.9 wt.% La, 18.50 wt.% Cr, 7.60 wt.% Ir and 18.97 wt.% O. The percentage of La, Cr, Ir and O obtained from the EDX analysis are 53.7, 20.6, 6.8 and 18.9 wt.% respectively. On the other hand, the percentages of La, Cr, Ir and O obtained from ICP-OES analysis are 53.6, 20.7, 7.3 and 18.4 wt.%, respectively. The TEM micrograph in **Fig. 5.17** reveals crystalline particles resembling a quasi-spherical primary particle with an average particle size of ~200 nm.

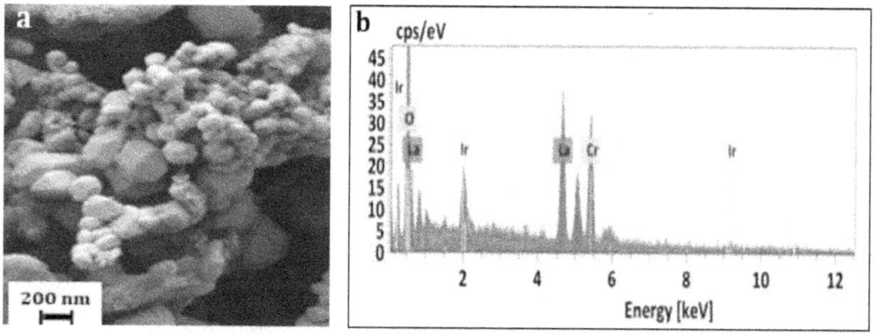

Figure 5.16 (a) SEM image and (b) EDX of LCIO-AV.

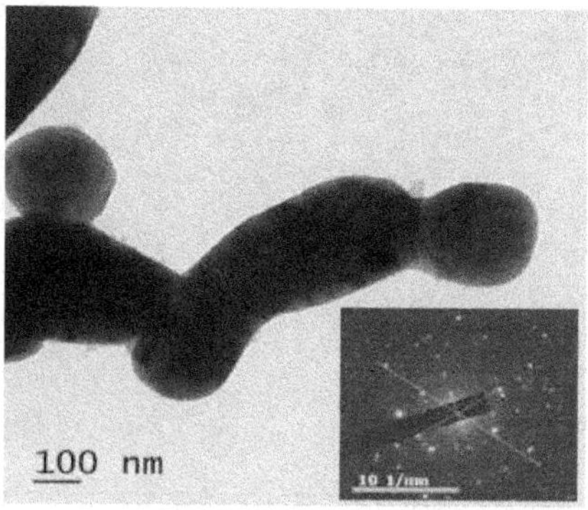

Figure 5.17 TEM image of LCIO-AV.

5.3.2.1 Catalytic evaluation of LCIO-AV

The catalytic activity of LCIO-AV toward the decomposition of hydroxylammonium nitrate (HAN) is evaluated using TGA, differential scanning calorimetry (DSC) and batch reactor experiments. In TGA, LCIO-AV exhibits an onset temperature of 114 °C and a peak decomposition temperature of 127 °C. The reduction in both onset and peak decomposition temperatures in the presence of LCIO-AV suggests enhanced catalytic activity. The TG/DTG plot of HAN without and with LCIO-AV catalyst is shown in **Fig. 5.18**. The onset decomposition temperature achieved with LCIO-AV is slightly lower than that observed for the Co-substituted lanthanum chromite system.

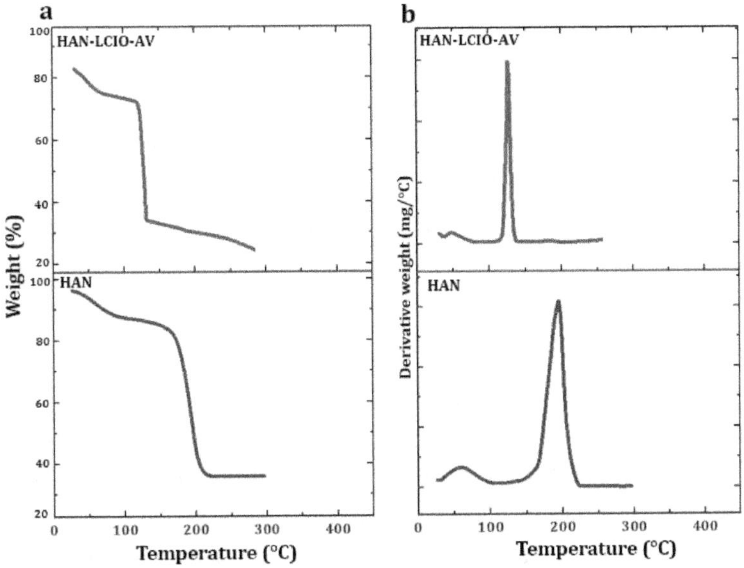

Figure 5.18 (a) TG plot and (b) DTG plot of HAN with LCIO-AV and without LCIO-AV catalyst

DSC analysis provides insight into the thermal decomposition behaviour of HAN in the presence of the catalyst and allows assessment of the catalytic efficiency through changes in decomposition temperature. DSC plot of HAN and HAN-LCIO-AV is given in **Fig. 5.19**. The DSC exhibits a pronounced exothermic peak corresponding to the catalytic decomposition of HAN. The onset of decomposition (Tonset) is observed at approximately 130 °C, with the peak decomposition temperature (Tpeak) occurring at 132°C relative to a Tonset of 149 °C and Tpeak of 152 °C for HAN without a catalyst. These results indicate effective catalytic activation of HAN decomposition by the LCIO-AV catalyst.

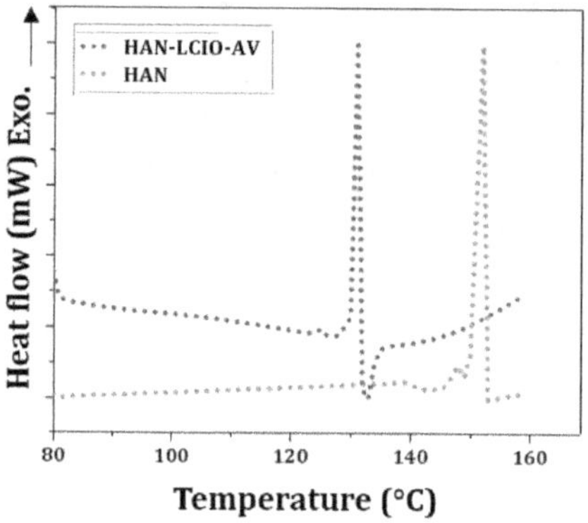

Figure 5.19 DSC plot of HAN and HAN with LCIO-AV

The performance of LCIO-AV is further evaluated using a batch reactor, where pressure fluctuations during the decomposition of the propellant are monitored in pulse mode operation. HAN on decomposition with LCIO-AV gives a pressure rise of 1.9 bar. The pressure increase obtained indicates a more efficient catalytic reaction and a higher decomposition rate. The pressure-time curve obtained for HAN-LCIO-AV is shown in **Fig. 5.20**. In addition, the ignition delay associated with HAN decomposition using the LCIO-AV catalyst is found to be in the range of 0.2–0.3 s, which is slightly lower than that observed for the LCCO-AV catalyst, which exhibits an ignition delay time of 0.4 s. The reduced ignition delay highlights the superior catalytic activity of LCIO-AV, enabling faster initiation of HAN decomposition. Performance comparison of HAN with and without the catalysts in terms of onset and peak decomposition temperatures and ignition delay is provided in **Table 5.3**. Even though the peak decomposition temperature observed in the present study is higher than the reported value using Ir/γ-Al_2O_3 catalyst (Agnihotri and Oommen, 2020), the LCIO are promising for HAN-based propulsion system due low cost and high temperature stability.

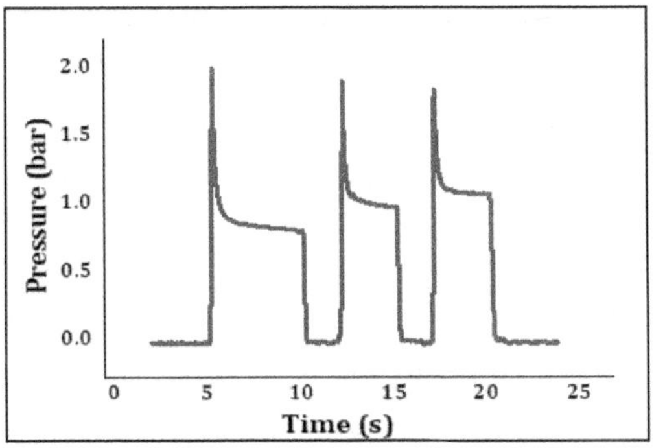

Figure 5.20 Pressure-time graph from the batch reactor test of HAN-LCIO-AV.

Table. 5.3 Performance comparison of HAN with and without catalyst from TG/DTA and batch reactor test.

System	Tonset (°C)	Tpeak (°C)	Ignition delay (s)
HAN-LCIO-AV	114	127	0.2-0.3
HAN-LCCO-AV	119	141	0.4
HAN alone	158	195	14.5

The analysis of evolved gases using a mass spectrometer coupled with the TG-MS reveals peaks corresponding to the molecular mass (m/z) of 18, 30, 44, and 28, indicating the formation of H_2O, NO, N_2O and N_2 molecules on decomposition of HAN in the presence of LCIO-AV. The evolved gas analysis result of HAN decomposition in presence of LCIO-AV catalyst is shown in **Fig. 5.21**.

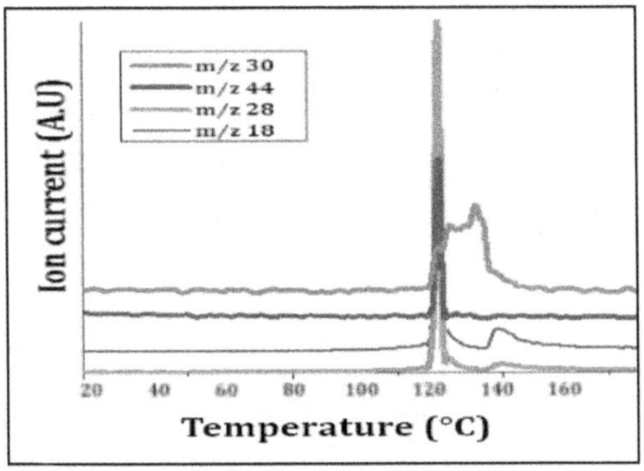

Figure 5.21 Evolved gases during the decomposition of HAN with LCIO-AV catalyst.

The surface area evaluated using the BET adsorption isotherm analysis indicates that LCIO-AV possesses a specific surface area of 3.9 m^2/g. In spite of the lower surface area of the catalyst, the observed catalytic activity can be governed by the Ir active sites on the surface of LCIO-AV, facilitating HAN decomposition.

Ir in LCIO exist as Ir^{4+} and introduces an excess positive charge into the lattice. In order to maintain electroneutrality, charge compensation occurs by the creation of cation vacancy or adsorption of O^-/OH^- and O^{2-} onto the anion vacant sites on the surface. This cation vacancy enhances the catalytic activity by increasing the active sites, modulating the electron density and accelerating the charge transfer. Since doping can lead to distortion of the octahedra, affect the Cr-O orbital overlapping, and facilitate lattice oxygen mobility. Hence, the lattice oxygen can migrate from the bulk and occupy the oxygen vacancies, which act as active sites for the reaction. The lattice oxygen and surface O^- / OH^- groups participate in the redox reactions with HAN.

The adsorption of the HAN molecule on catalyst surface through either hydrogen bonding or coordination of one of the hydrogen atoms from hydroxylamine to the oxygen atom or hydroxyl group of the catalyst explain the mechanism of HAN decomposition. This interaction facilitates the initial proton transfer reaction, resulting in the formation of hydroxylamine and HNO_3, thereby lowering the activation energy for decomposition. Subsequent reactions between hydroxylamine and nitric acid yield N_2, and oxides of nitrogen (Tian et al., 2024; Vargeese and Muralidharan, 2012). In addition, the Ir^{4+} in the LCIO lattice facilitate the HAN decomposition through the Ir^{4+}/ Ir^{3+} redox mechanism (Eq. (5.9) and (5.10)). These features may be attributed to the enhanced catalytic activity of LCIO-AV, as evidenced by the lower decomposition temperature observed during thermal analysis.

$$Ir^{4+} + NH_3OH^+ \rightarrow Ir^{3+} + \tfrac{1}{2} N_2 + 2H^+ + H_2O \qquad (5.9)$$

$$Ir^{3+} + 3/2\ H^+ + \tfrac{1}{2} NO_3^- \rightarrow Ir^{4+} + \tfrac{1}{2} HNO_2 + \tfrac{1}{2} H_2O \qquad (5.10)$$

5.4 Conclusions

Catalytic decomposition is the primary propulsion mechanism in monopropellant-based satellite thrusters. In this work, transition-metal-substituted lanthanum chromite perovskites ($LaCr_{1-x}M_xO_3$, x = 0-0.5, M= Co or Ir) are employed as catalysts owing to their thermal stability, mechanical strength, high melting point (~2500 °C), and catalytic activity. Cobalt-substituted lanthanum chromite perovskite powder with optimal composition of $LaCr_{0.8}Co_{0.2}O_3$ is prepared by the solution combustion method with (LCCO-AV) and without (LCCO) aloe vera gel capping agent using urea as a fuel. The onset and peak decomposition temperatures of HAN is lowered by 39 °C (158 to 119 °C) and 54 °C (195 to 141 °C), respectively, in the presence of LCCO-AV catalyst. The presence of cobalt in variable oxidation state influences catalytic activity. The performance evaluation of the LCCO-AV using a batch reactor test setup, in pulse mode operation exhibits a pressure increase of 1.7 bar with an ignition delay time of 0.4 s. Ir substituted lanthanum chromite with an optimal composition of $LaCr_{0.90}Co_{0.1}O_3$ exhibits better catalytic activity with a further lowering of the onset and peak decomposition temperatures to 114 °C and 127 °C, respectively. A pressure spike with an ignition delay of 0.2-0.3 s is observed in batch-reactor tests using LCIO-AV catalyst. The developed LCCO-AV and LCIO-AV provide a promising alternative for the costly Pt and Ir based catalyst system for HAN based monopropellant. The development of these catalysts can pave the way to replace the currently used toxic hydrazine monopropellant with the green alternative, HAN.

CHAPTER 6

CONCLUSIONS AND FUTURE DIRECTIONS

6.1 Conclusions

Hydrazine is the state-of-the-art monopropellant with a rich heritage of more than half a century owing to its reliable performance characteristics. However, its toxicity and high vapour pressure increase the cost of satellite launches due to the expensive propellant handling requirements. Hence, the development of a green monopropellant is highly beneficial, greatly reducing the hazards associated with handling and safety. Among different monopropellants, hydroxyl ammonium nitrate (HAN) is the most promising due to its negligible vapor pressure, low freezing point, higher density (1.8 g/cc) and higher density impulse (470 $\times 10^3 kg.s/m^3$ for 75% HAN). For propellant applications, a fuel component is added to HAN solution to enhance its performance. The added advantages of fuel in the system are its ability to improve the combustion process, control the burning rate, reduce thermal stress and minimise structural degradation. The fuel component can be either molecular or ionic liquids (IL). ILs possess several superior characteristics such as negligible vapour pressure and wide operating temperature that make them suitable for space technology applications. Moreover, the designability of the ILs by the choice of cations/ anions, functional group, or even the length of the alkyl chain can tailor their properties. The challenge is to find a suitable IL with high energy density, positive heat of formation, high density, compatibility and solubility to formulate a HAN-based monopropellant.

A brief review of hydroxyl ammonium nitrate-based propellants is given in **Chapter 1.** This chapter summarises the synthesis methods and thermal and catalytic decomposition behaviour of HAN. The catalyst development, and fuel formulation of HAN based propellants are also described. Most of the high performing HAN formulations are based on methanol as fuel because of its favourable combustion characteristics and improved specific impulse. However ionic liquid fuels effectively address the limitations such as volatility, toxicity, and handling hazards associated with conventional molecular fuels. They offer several inherent advantages, including high thermal and chemical stability, and negligible vapour pressure. The high heat of formation of IL fuel contributes to high combustion chamber temperature and eventually high specific impulse.

The synthesis of hydroxylammonium nitrate (HAN), by the neutralization method and its concentration to 80wt. % by vacuum distillation for propellant application is discussed in **Chapter 2.** The purity and concentration of HAN are determined by chemical analysis and spectroscopic methods. The 80 wt.% HAN used throughout this study has a density of 1.57778 g/cc at 20 °C. The HAN is thermally stable up to 158°C with an exothermic decomposition at Tpeak of 195°C. The evolved gas analysis using mass spectrum shows the formation of safe combustion products such as NO, N_2O and N_2. The batch reactor test in pulse mode operations gives a pressure rise to the maximum value of 0.9 bar with an ignition delay of 14.5 s. Theoretical studies using NASA-CEA indicate that 80% HAN yields a specific impulse ($Isp_{(vac)}$) of about 205 s.

The application of imidazolium cation-based ionic liquid fuels in HAN formulations is studied in **Chapter 3.** Density Functional Theory (DFT) is used to screen ionic liquids based on 1-butyl-3-methyl imidazolium (BMIM) cation and five energetic anions viz., nitrate (N), dinitramide (DN), 3-amino-1,2,4-triazolate (Tz), 3,5-diamino-1,2,4-triazolate (DTz) and 5-amino tetrazolate (AT). The computational insights help to identify promising energetic fuel candidates. 1-butyl-3-methyl imidazolium-5-amino tetrazolate (BMIMAT) is selected as the candidate fuel due to the high heat of formation and is synthesised via a metathesis reaction. Thermal behaviour of BMIMAT is assessed using DSC/TGA and Py–GC–MS. The S_N2 mechanism for thermal decomposition of BMIMAT is proposed based on the mass spectral analysis, which is complemented by theoretical studies. HAN-BMIMAT system delivers a maximum theoretical $Isp_{(vac)}$ of 324 s at an O/F ratio of 6. BMIMAT is soluble in 80wt.% HAN solution, enabling the preparation of monopropellant formulations by simple mixing. The thermal evaluation of HAN-BMIMAT at an O/F ratio of 6 shows an onset decomposition temperature of 135 °C and a peak decomposition temperature of 146 °C. The reduced onset decomposition temperature compared to that of pure HAN and BMIMAT indicates enhanced energetic characteristics arising from synergistic interactions between HAN and the ionic liquid fuel.

Chapter 4 evaluates the effect of variation of side chain (ethyl, butyl, allyl, vinyl) in the imidazolium cation on the energy content, thermal stability, and reactivity of ionic liquids paired with energetic anions. DFT predicted HOMO-LUMO energy gaps and heats of formation values are compared to identify optimal IL fuel. The vinyl-substituted cation paired with the 5-amino tetrazolate (AT) anion (VMIMAT) is highly energetic owing to its high heat of formation, but the low HOMO-LUMO energy gap (2.8 eV) indicates increased sensitivity and reduced stability. However, the allyl-substituted AMIMAT exhibits a high heat of formation (485 kJ/mol) while offering improved stability (HOMO-LUMO energy gap ~3.6 eV), making it a more promising fuel candidate. AMIMAT, a room-temperature ionic liquid (RTIL) is synthesised by metathesis reaction for the first time. The onset and peak decomposition temperatures observed for AMIMAT are 180 °C and 256 °C respectively, and the E_a values for the thermal decomposition is ~121 kJ/mol. The HAN-AMIMAT system delivers a specific impulse of 324 s at an O/F ratio of 5, which is significantly higher than that of 95% HAN. HAN–AMIMAT formulations are prepared at different oxidizer-to-fuel (O/F) ratios, and their thermal behaviour is studied. At lower O/F ratios of 1 to 4, precipitation is observed due to the low solubility of the double salt resulting from the interaction between hydroxylammonium nitrate and AMIMAT which is confirmed by theoretical studies. AMIM---AT interaction is weakened, as evidenced by the reduction in the BCP electron density from 0.0524 au to 0.02139 au upon interaction with HAN. This strong ion pairing reduces the solubility of the resulting complex at lower oxidizer-to-fuel ratios, leading to the observed precipitation. HAN-AMIMAT formulation corresponding to the theoretically optimized O/F ratio of 5, exhibits an onset decomposition temperature of 120 °C and a peak decomposition temperature of 147 °C.

The decomposition of hydroxylammonium nitrate (HAN) can proceed through either thermal or catalytic pathways. Catalytic decomposition is advantageous as it promotes highly exothermic reaction routes at significantly lower onset temperatures, leading to enhanced energy release. At present, noble metals such as platinum and iridium are predominantly employed as catalysts in aerospace propulsion systems. However, the development of low-cost alternative catalysts presents a promising and economically viable substitute to expensive Pt- and Ir-based catalyst systems for HAN-based monopropellants. **Chapter 5** describes the development of perovskite catalysts for the thermal decomposition of HAN. Cobalt and Iridium substituted lanthanum chromite perovskite powders are prepared by the solution combustion method using urea as a fuel and aloe vera gel as a capping agent. The composition of Co substituted and Ir substituted lanthanum chromite perovskite catalysts is optimised as $LaCr_{0.8}Co_{0.2}O_3$ and $LaCr_{0.9}Co_{0.1}O_3$, respectively. The catalyst prepared in the presence of aloe vera gel capping agent exhibits lower particle size, spherical morphology and improved catalytic activity towards HAN decomposition compared to the one prepared without the aloe vera gel. The onset decomposition temperature and the peak decomposition temperature of HAN are lowered by 35 °C and 54 °C, respectively, in the presence of LCCO-AV ($LaCr_{0.8}Co_{0.2}O_3$ prepared with the aloe vera gel capping agent). The presence of cobalt in variable oxidation state influences catalytic activity. The performance of the LCCO-AV evaluated with a batch reactor, in pulse mode operation exhibits a pressure increase of 1.7 bar with an ignition delay time of 0.4 s. The LCIO-AV ($LaCr_{0.9}Ir_{0.1}O_3$ prepared with the aloe vera gel capping agent) exhibits better catalytic activity with a further lowering of the onset temperature of decomposition to 114 °C and peak decomposition temperature to 127 °C, when compared to LCCO-AV, along with a lower ignition delay of 0.2–0.3 s in batch-reactor tests.

Major findings of the work are:

- The green monopropellant, HAN is synthesized and concentrated to 80 wt.%
- Theoretical performance evaluation of $HAN_{80\%}$ using NASA-CEA predicts a specific impulse ($Isp_{(vac)}$) of 205s which can be improved through the addition of suitable fuel
- Density Functional Theory (DFT) is used to screen ionic liquid fuels based on 1-butyl-3-methyl imidazolium (BMIM) cation paired with energetic anions
- Energetic ionic liquid fuel with BMIM cation and 5-amino tetrazolate (AT) anion exhibits the highest heat of formation

- BMIMAT is synthesized through a metathesis reaction
- HAN-BMIMAT delivers an $Isp_{(vac)}$ of 324 s at oxidizer to fuel ratio (O/F) of 6
- HAN-BMIMAT formulation exhibits enhanced energetic characteristics arising from synergistic interactions between HAN and BMIMAT
- The influence of the side chain in the cation on energy content for imidazolium ionic liquids is compared to identify optimal IL fuel based on DFT studies
- Allyl-substituted AMIM cation with 5-amino tetrazolate (AT) anion gives a higher positive heat of formation, along with better stability
- AMIMAT, a room-temperature ionic liquid (RTIL) is synthesised by metathesis reaction for the first time and its thermal behaviour is evaluated
- HAN-AMIMAT formulation corresponding to the theoretically optimized O/F = 5, can deliver an $Isp_{(vac)}$ of 324 s which is ~58% higher than with HAN alone.
- Theoretical studies indicate stronger interactions between HA^{+} and AT^{-} in HAN-AMIMAT formation, favouring the formation of double-salt
- The precipitate formation at O/F below 5 can be due to the lower solubility of the double salt
- Substituted lanthanum chromite, $LaCr_{0.8}Co_{0.2}O_3$ and $LaCr_{0.9}Ir_{0.1}O_3$, are promising and economically viable substitutes to expensive Pt- and Ir-based catalysts for HAN decomposition
- Use of aloe vera gel as a green capping agent improves the catalytic activity of substituted lanthanum chromite towards HAN decomposition
- Ir-doped lanthanum chromite (LCIO-AV) exhibits better catalytic activity and lower the onset and peak decomposition temperatures by 44 °C and 68 °C, respectively.
- LCIO-AV exhibited a decrease in combustion delay from 14.5 to ~0.3s, highlighting its potential for monopropellant catalyst applications

6.2 Future Directions

Imidazolium-based ionic liquid fuels, when paired with energetic anions, have demonstrated enhanced performance and have been investigated as potential fuel candidates through density functional theory (DFT) studies. The selection of fuels is guided by the highest theoretical heats of formation obtained from these calculations. An increase in heat of formation generally corresponds to improved propellant performance, attributed to the higher density of energetic N–N bonds. The heat of formation can be further increased by replacing the imidazolium cation with triazolium or tetrazolium cations paired with suitable energetic anions. Accordingly, the design and development of room-temperature ionic liquids based on triazolium and tetrazolium cations, possessing good thermal stability and reduced sensitivity, is proposed as a future work.

The performance of HAN based formulations depends on the composition of monopropellant, type and structure of catalyst bed and catalyst preheat temperature. Future work emphasises the design of cost-effective and shaped catalysts with high surface area and high temperature tolerance to enable sustained decomposition at lower preheat temperatures and under repeated firing conditions.

Future research includes catalyst bed design and compatibility studies with thruster materials. Long-duration firing tests and thermal cycling studies are required for the practical applicability of newly developed HAN-based formulations and catalysts in actual propulsion environments.

www.ingramcontent.com/pod-product-compliance
Lightning Source LLC
LaVergne TN
LVHW021736071126
851857LV00040B/1328
9798256289751